The Petroleum Industry

Crude Oil Exploration & Refining
(Handbook & guide)

ALSO, BY P K SINGH

Books

ACCIDENT AND INCIDENT INVESTIGATION
(A complete Training guide with Report Writing)

CRACKING SAFETY AND HSE JOB INTERVIEW
(Includes 200+ important Questions and Answers)

INTERVIEWER'S CHOICEST QUESTIONS
(Covers difficult Interview Questions and Answers

PROCESS SAFETY ENGINEER INTERVIEW GUIDE

FIRE WATCH, SAFETY FOREMAN, RIGGER AND OPERATOR INTERVIEW GUIDE

BEYOND SOLITUDE – Alone no More

CONVERSATIONS WITH CHILDREN
(The Missed Dialogues)

Please visit:
https://www.amazon.com/author/pksingh

The Petroleum Industry

Crude Oil Exploration & Refining

(Simplified for Easy Understanding)

· · · · · · · · · · · ·

P K Singh

Copyright © 2023 by P K Singh

The right of P K Singh to be identified as the Author of The Work has been asserted by him in accordance with The Copyright, Designs and Patents act 1988

First published by Amazon in December 2023

This is a basic guide to understand the Oil & Gas Exploration & Refining.

All rights reserved. No part of this book may be reproduced or used in any manner without written permission of the copyright owner except for the use of quotations in a book review.

For more information: Contact Amazon

Book design by P K Singh

ASIN: B0CQ6YJTRD (Paperback)
ASIN: B0CQ7ZQ6PY (e-Book)
ASIN: B0CQCZSWMG (Hardcover)
ISBN-13: 979-8832595658 (Paperback)

https://www.amazon.com/author/pksingh

DEDICATION

To My Dearest Wife, Son, and Daughter

This book is a testament to the love and support that each of you has brought into my life.

To my wife, your unwavering encouragement and understanding has been the cornerstone of my journey. Your love is the inspiration behind every word written here.

To my son, watching you grow into the incredible person you are fills my heart with pride.

To my daughter, your grace, intelligence, and warmth bring joy to my days. May this book serve as a reminder of the boundless possibilities that await you and the love that surrounds you at every step of your journey.

Together, you form the tapestry of my life, and this book is a small reflection of the gratitude and love I feel for each of you. Thank you for being my pillars of strength, my sources of joy, and my greatest blessings.

With all my love,

P K Singh

TABLE OF CONTENTS

DEDICATION IX
TABLE OF CONTENTS XI
TABLE OF FIGURES XVII
ACKNOWLEDGMENTS XIX
FOREWARD XXI
PREFACE XXIII

THE PETROLEUM INDUSTRY 1
    UPSTREAM, MIDSTREAM & DOWNSTREAM 3
        Upstream 3
        Mid-Stream 6
        Downstream: Refining, Distribution, & Beyond 9

CRUDE OIL EXPLORATION 17
    ORIGIN OF OIL & GAS 19
    TYPES OF CRUDE OIL 23
    CHARACTERISTICS OF CRUDE OIL 26
        API GRAVITY: 26
        SULFUR CONTENT: 26

GEOPHYSICS AND SEISMIC TECHNIQUES 29
    FOUNDATION OF GEOPHYSICS 29
    SEISMIC TECHNIQUES IN CRUDE OIL EXPLORATION 30
    APPLICATIONS OF SEISMIC TECHNIQUES IN EXPLORATION 31
    ADVANCED SEISMIC TECHNOLOGIES 32
    CHALLENGES AND INNOVATIONS IN SEISMIC EXPLORATION 32

DRILLING TECHNOLOGIES AND WELL CONSTRUCTION 37
    Cable Tool Drilling (Percussion system) 38
    Rotary-hydraulic system 39
    CONVENTIONAL ROTARY DRILLING 40
    ADVANCED DRILLING TECHNOLOGIES 41
    OFFSHORE DRILLING TECHNOLOGIES 42

DRILLING CHALLENGES	43
INNOVATIONS AND FUTURE TRENDS	43

RESERVOIR ENGINEERING AND MANAGEMENT 45

KEY CONCEPTS IN RESERVOIR ENGINEERING	46
CHALLENGES IN RESERVOIR ENGINEERING	47
INNOVATIONS IN RESERVOIR ENGINEERING	48
DATA ANALYTICS AND MACHINE LEARNING:	48

UPSTREAM OILFIELD FACILITIES 51

DRILLING RIGS	51
TYPES OF DRILLING RIGS	54
PRODUCTION PLATFORMS	57
SEPARATION FACILITIES	59
Design Considerations for Two-Phase Separators	*59*
Three-Phase Separators	*60*
Design Considerations for Three-Phase Separators	*61*
Applications and Challenges	*61*
Innovations and Future Trends	*62*

OVERVIEW OF REFINING PROCESS 67

1. DISTILLATION: THE INITIAL SEPARATION PROCESS	67
2. CONVERSION PROCESSES: (ENHANCING PRODUCT YIELDS)	68
3. TREATMENT PROCESSES: IMPROVING PRODUCT QUALITY	69
4. SEPARATION PROCESSES: REFINING SPECIFIC PRODUCT STREAMS	70
5. FINAL BLENDING: ACHIEVING DESIRED PRODUCT SPECIFICATIONS	70
6. ENVIRONMENTAL CONTROLS: MITIGATING ENVIRONMENTAL IMPACT	71
7. PRODUCT DISTRIBUTION: DELIVERING TO END USERS	72

TYPES OF REFINERIES 75

SKIMMING PLANTS	75
LUBRICATING & WAX PLANTS	77
Lubricating Plants:	*77*
Wax Plants:	*78*
COMPLETE RUN-DOWN REFINERY	81
TOPPING REFINERY	84
HYDROSKIMMING REFINERY	85

    CONVERSION REFINERY 86
    DEEP CONVERSION REFINERY 88
        Key Processes in Deep Conversion: *89*
        1. Hydrocracking: *89*
        2. Visbreaking: *89*
        3. Coking: *89*
        4. Residue Desulfurization: *89*
        Significance of Deep Conversion: *89*
        Examples of Deep Conversion Refineries: *90*

BASIC CHEMISTRY 93

    HYDROCARBONS 93
    NON-HYDROCARBONS 94
    HYDROCARBON REACTIONS 94
        CRACKING: *94*
        DEHYDROGENATION: *95*
        HYDROGENATION: *95*
        PYROLYSIS: *95*
        CYCLIZATION: *96*
        ALKYLATION: *96*
        ISOMERIZATION: *97*
        POLYMERIZATION or COPOLYMERIZATION: *97*
        OXIDATION: *97*
        ETHERIFICATION: *98*
        HYDRODESULFURIZATION: *98*
        CATALYSIS: *98*

DISTILLATION AND FRACTIONATION 99

    PHYSICAL SEPARATION: 100
        DISTILLATION: *100*
        CRYSTALLIZATION: *108*
        ABSORPTION: *108*
        SOLVENT EXTRACTION: *108*
    CHEMICAL CONVERSION 109
        C/H RATIO: *109*

CATALYTIC CRACKING AND HYDROCRACKING: 111

THERMAL CRACKING:	111
CATALYTIC CRACKING:	112
FLUID CATALYTIC CRACKING (FCC):	112
HYDROCRACKING:	113
VISBREAKING:	114
COKING:	114
REFORMING:	115
THERMAL REFORMING:	116
CATALYTIC REFORMING:	116
PROPANE DEASPHALTING:	117
TREATING PROCESSES	**118**
HYDROTREATING:	118
HYDRO-REFINING:	119
CRUDE DESALTING:	119
SULFURIC ACID TREATMENT:	122
CLAY REFINING:	123
CAUSTIC WASHING:	123
SWEETENING:	123
ASPHALT BLOWING:	123
DOCTOR TREATMENT:	124
HYPOCHLORITE TREATMENT:	124
MEROX PROCESS:	124
AFTER TREATMENT:	124
OTHER PROCESSES	**125**
BLENDING:	125
BLENDSTOCK CHARACTERISTICS:	127
EMULSIFYING:	128
GREASE MAKING:	128
SULFUR RECOVERY:	128
BLOWDOWN SYSTEM:	128
PROCESS HEATERS:	129
FUGITIVE EMISSIONS:	129
UTILITIES AND SUPPORT OPERATIONS:	129

DESULFURIZATION & ENVIRONMENTAL COMPLIANCE 131

Understanding Desulfurization 131
The Sulfur Problem: 131

Desulfurization Techniques:	*131*
Environmental Impact:	*132*
Global Regulatory Landscape:	*132*
CHALLENGES IN DESULFURIZATION	133
Technological Challenges:	*133*
Energy Efficiency:	*134*
INNOVATIONS AND FUTURE TRENDS	134
Emerging Technologies:	*134*
Hybrid Approaches:	*134*
PETROCHEMICAL INTEGRATION AND DOWNSTREAM PRODUCTS	**135**
MEASUREMENT OF OIL & GAS	**137**
CRUDE OIL PRICING	**139**
TAXATION ON PETROLEUM PRODUCTS	**141**
1. Federal Excise Tax:	*141*
2. State Taxes:	*142*
UNITED KINGDOM (UK):	142
1. Fuel Duty:	*142*
2. Value-Added Tax (VAT):	*142*
3. Carbon Price Floor:	*142*
INTERNATIONAL CONSIDERATIONS:	143
1. Cross-Border Fuel Pricing:	*143*
2. Harmonization Efforts:	*143*
3. Environmental Taxes:	*143*
EMERGING TECHNOLOGIES AND FUTURE TRENDS	**146**
INTERNATIONAL SCENARIO (2022-23)	**157**
BRIEF SUMMARY	**159**
BASIC PROCESSING OPERATIONS	159
PHYSICAL & CHEMICAL PROCESSES	160
ABOUT THE AUTHOR	**161**

TABLE OF FIGURES

Figure 1: Upstream - Offshore Drilling Rig _____ 3
Figure 2: Downstream Refinery _____ 9
Figure 3: Standard Cable Tool Drilling System _____ 38
Figure 4 : Drilling layers of rock & soil _____ 39
Figure 5: Onshore Drilling Rig _____ 54
Figure 6: Offshore Drilling Rig _____ 55
Figure 7: Skimming Plant (Refinery) _____ 75
Figure 8: Complete Run down Refinery _____ 80
Figure 9: Typical Crude Distillation & Downstream products _____ 82
Figure 10: Product comparison of 4 types of refineries _____ 83
Figure 11: A typical distillation column _____ 101
Figure 12: A typical Distillation _____ 102
Figure 13: Crude distillation unit (simplified) _____ 104
Figure 14: Single stage Electrostatic Desalting _____ 121
Figure 15: Two stage Electrostatic Desalting _____ 122
Figure 16: Blend stock Characteristics _____ 127

ACKNOWLEDGMENTS

I extend my deepest appreciation to my unwavering pillars of support throughout this literary journey. To my loving wife, whose boundless patience and encouragement provided the steady foundation upon which this book stands. Your belief in me and the countless sacrifices you made did not go unnoticed, and I am profoundly grateful.

To my son, whose enthusiasm and curiosity added a unique perspective to the writing process. Your shared excitement for the subject matter breathed life into these pages, and your presence brought joy to the long hours of crafting and editing.

A special acknowledgment goes to my daughter, whose insightful feedback, creativity, and unwavering belief in the importance of storytelling infused this book with a sense of purpose. Your contribution, both big and small, made this journey richer and more meaningful.

To my entire family, thank you for the understanding during those solitary writing hours and for being a constant source of inspiration. Your love and support made turning the pages of this chapter all the more rewarding.

This book is a testament to the strength of our familial bonds, and I dedicate its pages to the cherished moments we've shared and the countless more that await.

With sincere appreciation,

P K Singh

FOREWARD

This book is written with the purpose of helping the reader, far from the Oil Fields, to "understand" the Petroleum Industry—as it really is.

Millions of people who are interested either directly or indirectly in the Petroleum Industry may never get a chance to see an oil well, or a Petroleum Installation tank farm, or a pipe line, or a Refinery—but they want to know, and should know, about Petroleum. Petroleum and its derivative products are very important to us today.

Petroleum has been in usage since centuries as is known from the fact that Zoroastrians of Persia worshipped oil and the same is mentioned in history as "Fire Worshippers."

In the seventh century petroleum was known in Japan as "Burning Water." Chinese history mentions the usage of oil and gas for lighting purposes.

This book is an attempt to clarify and tell in simple terms the story of Petroleum industry.

PREFACE

This Brief book is a brief guide on Upstream Petroleum Industry, providing basic knowledge on Crude oil refining operations. The information is given in simple terms which are easy to understand.

Most of us are unaware of the Petroleum Industry working and we just think that Diesel, (Gas Oil) Petrol (Gas), Kerosene, Jet Fuel and Lubricants are all that are manufactured by Petroleum Industry.

Petroleum refining is one of the most important sectors in the complex web of modern society. With the world about to enter a new era marked by ever-growing energy demands and environmental concerns, it is critical that we grasp the subtleties of crude oil refining.

This book, delves deeply into the center of the petroleum refining operations. The goal is to simplify the difficult process of turning raw crude oil into the various products that power our daily existence. The refinement process is the alchemical crucible that transforms black gold into the fuel of the contemporary world, from gasoline that drives our cars to petrochemicals that form our materials and medicines.

This work began with the realization that many people still find petroleum refining to be mysterious, even though it plays a vital role in our daily lives. This book attempts to close the knowledge gap between the general public, who are keen to understand the amazing journey from crude oil to refined products, and the technical complexities of the refining sector.

We will explore the science, technology, and engineering that support the refining process as we set out on this adventure together. Every chapter will reveal a new level of intricacy, from hydro-treating and catalysis to distillation and cracking, giving readers a greater understanding of the skill required to transform a raw material into a wide range of necessary products.

Furthermore, this book recognizes the changing landscape of energy and environmental consciousness; it does not live in a vacuum. We shall explore the potential and difficulties the refining sector faces in the twenty-first century within its pages, including the continuous search of energy independence, sustainability, and cleaner technology.

My goal in writing this book is to inspire astonishment at the wonders of human ingenuity in creating the world we live in, in addition to imparting knowledge. Join me on this fascinating journey through the crucible of petroleum crude oil refining whether you are a student of chemical engineering, a professional in the energy sector, or just a curious mind eager to understand the inner workings of an industry that fuels our modern existence.

If we think beyond fuels for the automotive sector, liquid fuels have immense usage in creating overpasses, asphalt highways, parking lots, shopping malls and endless other uses.

Liquid fuels have helped massive container ships and trucks to bring essential and non-essential food items and other commodities from across the World to our country in shortest possible time all year round.

Modern life these days starts and ends amidst plastics, the manufacturing of which begins with feed stocks derived from petroleum. The medical equipment: injections, gloves, catheters, trays, basins, thermal blankets and lab ware all are essentially manufactured from petroleum.

A huge variety of plastics used in cars, trains, airplanes, offices, trains and factories are all derived from PVC (poly vinyl chloride) which is again produced from petroleum.

The oil rich countries, (whether it is US, Saudi Arabia, China, Russia or Canada) have empowered its rulers and dictators since ages. The oil ownership has given dictating powers to autocrats, financed terrorists and brought massive corruption (Nigeria, Russia, Malaysia or Indonesia).

In the early era, the basic fuel requirements were met from using Wood, coal, and crop residues (known as Biomass fuels). These were not renewable as it led to deforestation and excessive usage of crop residues.

With the development of refined oil products like Gasoil, Kerosene, Diesel & Fuel oil, the energy consumption pattern changed drastically. The new fuels were cleaner, safer to produce, convenient to burn and were more energy efficient. By 1924 in road transport Diesel started replacing gasoline fueled vehicles and by 1930 most of the new trucks were powered by diesel engines. In 1936 the first saloon car with diesel engine was developed by Daimler Benz and became a favorite as Taxica

Diesel used as fuel is slightly cheaper than Gasoline and has 11% more energy than Gasoline in the same volume. Diesel is less flammable than Gasoline which makes it suitable in settings where fire is envisaged as an instant disaster (such as on-board ships). The early Diesel engines were too bulky to be used in automobiles. The first American Gasoline fueled car was built in 1892 by Charles Duryea. It was expensive and unreliable, and the car revolution came in 1908 by Henry Ford's introduction of car's "Model-T"

The development of improved cars started well after first World War in the United States of America, and then after 1950 in Japan and Europe and now throughout most of the countries worldwide.

Thermal efficiency of best cars is over 30% but engines in everyday use achieve near 25% only (friction losses and transmission losses account for major reduction in efficiency). Hybrid vehicles are much more efficient, and now we have the electric vehicles such as "TESLA."

In the aviation sector where Jet Fuel (Jet A and Jet A-1) are used, the engines have been undergoing improvements and new Boeing 787 (Dreamliner) is 70% more efficient than Boeing 707 Turbojet.

Crude oil refining is a key source of Petrochemical feedstock that is used to produce wide variety of synthetic materials. There are 2 kinds of feed stocks:-

a) Olefins – mainly Ethylene and Propylene

b) Aromatics – mainly Benzene, Toluene and Xylene

Ethylene is produced by steam cracking of naphtha and is most important feedstock. Polymerization of basic feedstock gives thermoplastics which account for more than 70% of polymers produced in the world.

The details of this unique Industry shall be uncovered more and more in simple terms in next chapters.

Chapter One

THE PETROLEUM INDUSTRY

In the ever-evolving landscape of the energy sector, knowledge is the key that unlocks the vast potential of one of the world's most critical resources – crude oil. This book is not just a handbook; it's a transformative guide that takes you on an illuminating journey through the intricacies of crude oil exploration and the refining process. This comprehensive compendium is meticulously crafted to empower professionals, students, and enthusiasts alike with a profound understanding of the complexities that govern the extraction and transformation of this liquid gold.

Commonly called "The Oil Industry" or more specifically referred to as Oil and Natural Gas Industry, it meets the worldwide requirement of fuel oils for automotive vehicles and raw material requirements of various processes of major Petrochemical based industries.

As of 2023, the United States remains the top producer of oil in the world, averaging 12.7 million barrels per day. Second and third place go to Saudi Arabia and Russia, with daily production of approximately 11 and 10.9 million barrels of oil, respectively. While natural gas production has risen from 54 billion cubic feet per day to more than 100 billion cubic feet per day.

While most refineries in other regions of the world optimize the production of diesel and jet fuel, most refineries in North America focus primarily on producing gasoline.

2 THE PETROLEUM INDUSTRY – CRUDE OIL EXTRACTION AND REFINING

According to figures from 2021, the petroleum sector generates one of the highest annual revenues, anticipated to be above $3 trillion.

According to the Energy Information Administration, the United States has been the world's greatest crude oil production since 2018, surpassing Russia and Saudi Arabia. The Energy Information Administration (EIA) predicts that US crude oil production will reach new highs in 2023 and 2024.

Petroleum refineries are sophisticated, capital-intensive facilities that use main and secondary processes, such as blending, to transform crude oil into more than 2600 finished products. Among the products in the assortment are:

LPG (liquefied petroleum gas)	Gasoline
Jet Fuel (Aviation Turbine Fuel)	Diesel
Kerosene	Fuel oil
Lubricating oil /wax	Petrochemical feed stock
Asphalt	Heating oil

Out of above list the transportation fuels are most valuable and in great demand.

UPSTREAM, MIDSTREAM & DOWNSTREAM

In broad sense of classification of Petroleum Industry, it has 3 major divisions:
1. Upstream (exploration and production)
2. Midstream (transportation and processing)
3. Downstream (distribution and sales to end users and consumers)

Some of the companies are only Upstream, some are only Downstream, while some are having both divisions. Let us understand more about the Upstream, Mid-stream and Downstream companies.

Figure 1: Upstream - Offshore Drilling Rig

UPSTREAM

The upstream sector of the oil and gas industry is the first stage of the energy value chain, covering activities such as crude oil and natural gas exploration, drilling, and production. Because it entails discovering, extracting, and bringing these rich resources to the surface, this critical section serves as the cornerstone for the entire business. Understanding the intricacies and dynamics of the upstream oil sector reveals the intricate procedures involved in extracting the Earth's subsurface treasure.

The Upstream companies manage the exploration and production of petroleum products (also called E & P companies).

Seismic Surveys and Geophysical Exploration: Upstream oil exploration begins with a thorough grasp of subsurface geology. Seismic surveys, which use sound waves to provide comprehensive photographs of geological features beneath the Earth's surface, are critical in locating potential reservoirs. Geophysicists examine the data to identify interesting areas for future exploration.

Exploratory drilling, which is frequently preceded by the drilling of wildcat wells, seeks to validate the presence of hydrocarbons in the targeted formations. Cores retrieved from these wells provide vital geological information, allowing geoscientists to appraise the discovery's potential. If the well proves to be commercially successful, it will serve as the cornerstone for future growth.

Drilling Types:

Exploratory Drilling: As previously said, exploratory drilling is the first phase aimed at locating new reserves.

Appraisal drilling is done after a discovery to determine the size and commercial feasibility of the reservoir. To define the boundaries of the discovery, many wells may be dug.

Development drilling happens once a field is judged commercially viable in order to maximize recovery. To tap into the reservoir's potential, many wells are deliberately dug.

UPSTREAM OPERATIONS CHALLENGES:

Complexity of Reservoirs: Some reservoirs present geological obstacles, such as complex structures or unconventional formations, necessitating specialized drilling and extraction procedures.

Concerns about the environment: Upstream operations are being scrutinized for their environmental impact, which includes potential spills, pollutants, and habitat destruction.

Technology Innovation: To combat obstacles, the sector consistently invests in technology innovation, from creating better drilling techniques to enhancing environmental performance.

Some of the upstream oil companies are listed below:

- Sinopec, China
- Petro China, China
- Chevron
- Shell
- ExxonMobil

- Philips 66
- ONGC
- Oil India

There are five phases in the life cycle of upstream oil and gas industry:
1. Explore
2. Appraisal
3. Develop
4. Produce
5. Close

1. **Explore:** (This phase usually lasts for 1 to 5 years)

Geological surveys are carried out to explore the potentially viable Oil/Gas sources. The government grants access to firms to explore via bidding process or direct awarding of contracts. Concessions are given to international companies for carrying out exploration in designated areas with contracts governing the rights of any oil/gas discovered.

Often it is seen that no viable oil/gas sources are discovered and the operations are terminated.

If viable sources are identified then further exploration occurs and companies plan for next phase. Social and economic studies are conducted for environmental impact assessments.

2. **Appraisal:** (This phase usually takes 4 to 10 years)

Sites are examined in more detail, drilling is planned, site infrastructure is developed and oil/gas reserves are mapped.

3. **Develop:** (This phase usually takes 4 to 10 years)

The site is prepared for production, Government permits are revised and updated and oil/gas is produced at the end of this phase.

4. **Produce:** (This phase usually lasts for 20 to 50 years)

Oil and gas are extracted and transported for processing and distribution. The output volume is variable and reduces towards the end of this phase.

5. **Close:** (This phase usually takes 2 to 10 years)

When it is observed that the exploration site is no longer cost effective to extract the remaining reserve, the site is decommissioned. It is the responsibility of the operating companies to return the site to as close as the original state as possible. This phase can take more time if environmental monitoring parameters need to be maintained.

It may be noted that majority of the projects do not reach steps 3, 4 & 5 because the reserves may not be sufficient to justify the amount of investment needed to carry on the extraction.

MID-STREAM

The midstream oil business is the crucial connection that connects the upstream and downstream segments of the energy supply chain. The upstream sector searches for and extracts crude oil and natural gas, while the downstream sector refines and distributes finished goods to customers. Transporting, storing, and marketing crude oil and its derivatives at wholesale prices are all part of midstream operations, which are vital to maintaining a smooth flow of energy resources from production hubs to end users.

The Mid-stream companies are engaged in transportation of raw crude from the wells to the refineries to extract oil and gas. In some cases, processing may happen at the production stage as is the case in LNG (Liquefied Natural Gas) and it happens prior to transportation. These companies have a good network of trucks, pipelines, shipping and storage facilities which is the key requirement to facilitate the operation.

IMPORTANT ELEMENTS OF THE MIDSTREAM OIL SECTOR:

1. Pipelines: Acting as the arteries for the long-distance transportation of natural gas and crude oil, pipelines constitute the foundation of midstream infrastructure. These complex networks of pipelines span continents, linking production sites and oil fields to refineries, distribution hubs, and eventually, markets. Pipelines are a popular means of transportation because of their efficiency and affordability, which guarantee a steady and dependable flow of hydrocarbons.

2. Rail and Truck Transport: Midstream companies use rail and truck transport to carry crude oil and processed products in areas where pipelines may not be feasible or sufficient. Because of its adaptability, transportation can be provided to places without pipeline infrastructure or in reaction to demand fluctuations. Tanker tankers and railcars function as alternate means of transport for oil.

3. Storage Facilities:

Facilities for strategic storage are essential to maintaining the stability of the energy supply chain. Crude oil and refined products are stored for both short and extended periods of time in terminals, tanks, and caverns. This tactical reservoir makes it possible to balance variations in supply and demand, guaranteeing consumers a consistent and dependable supply even in times of peak demand or production interruptions.

4. Transportation by Sea:

Marine transportation becomes an essential part of midstream operations for areas with access to waterways and for international trade. Energy resources are traded globally through the transportation of crude oil and processed goods across oceans by tankers and barges. In the global midstream network, ports with capabilities for loading and unloading cargo are vital nodes.

THE MIDSTREAM SECTOR'S IMPORTANCE:

1. Bridging Upstream and Downstream: The midstream sector serves as a vital link between the upstream and downstream parts of the oil and gas industry. It provides a smooth transition from crude oil extraction to refined product delivery by enabling the transport of hydrocarbons from production sites to refineries and, eventually, distribution terminals.

2. Ensuring Energy Security: Midstream infrastructure helps to ensure energy security by establishing a resilient and redundant network for delivering and storing crude oil and processed products. Diverse modes of transportation, storage facilities, and international shipping capabilities improve the industry's ability to adjust to interruptions, geopolitical shifts, or unforeseen occurrences affecting energy supply.

3. Market Access and Global Trade: Midstream activities provide access to a variety of markets while also promoting global trade in energy resources. Pipelines, rail transport, and sea shipping connect energy-rich regions with high-demand regions, allowing economic growth and guaranteeing a more equal worldwide distribution of resources.

4. Environmental and Safety Considerations: Because the midstream sector is so important in the transportation of hazardous commodities, environmental and safety concerns are vital. To limit environmental damage, prevent spills, and protect the safety of both workers and the communities through which these transportation routes pass, strict rules and industry standards control the design, building, and operation of midstream infrastructure.

INNOVATIONS AND CHALLENGES:

1. Aging Infrastructure:

The aging infrastructure, particularly in places with long-established networks, is one of the difficulties confronting the midstream business. Pipelines, storage tanks, and transport fleets must be upgraded and maintained to ensure the continuous reliability and safety of midstream activities.

2. Integration of Technology:

Innovation and technology are becoming increasingly important in optimizing midstream operations. The combination of modern sensors, data analytics, and automation improves midstream infrastructure monitoring, maintenance, and efficiency. In the complex midstream environment, smart technologies provide real-time data analysis, predictive maintenance, and improved decision-making.

3. Regulatory Compliance:

The midstream industry operates in a highly regulated environment with stringent compliance standards for safety, environmental impact, and community participation. Midstream operators have an ongoing struggle in navigating these regulatory frameworks while maintaining operational efficiency.

The midstream oil industry is the unsung hero of the global energy supply chain, enabling the seamless and efficient movement of energy resources from extraction to consumption. From pipelines crossing vast landscapes to tankers sailing the open seas, the midstream sector's infrastructure serves as the energy ecosystem's circulatory system.

THE PETROLEUM INDUSTRY 9

Figure 2: Downstream Refinery

DOWNSTREAM: REFINING, DISTRIBUTION, & BEYOND

The downstream oil sector is the engine that drives modern society, producing the fuels that move vehicles, planes, and ships, as well as the raw materials for numerous consumer items. Its influence extends across various industries, from transportation to manufacturing, and its evolution is driven by a relentless pursuit of efficiency, sustainability, and technological advancement.

The downstream refineries are responsible for removing the impurities from crude oil and converting the crude oil into usable products like gasoline, naphtha, diesel, kerosene, fuel oil, jet fuel, heating oil and asphalt.

The last stop on the energy trip is the downstream sector of the oil and gas industry, where crude oil is refined into a variety of valuable goods that power our everyday life. This complex phase entails converting crude oil into products and fuels that can be used, supplying them to consumers, and carrying out a number of industrial operations to produce a wide variety of items based on petrochemicals.

The downstream oil sector is essential to contemporary economies and serves as a catalyst for many other businesses in addition to producing energy.

REFINING:

1. Crude Oil Refining: In the downstream oil sector, refining involves exposing crude oil to a number of intricate procedures in order to reveal its hidden benefits. It is comparable to alchemy. The atmospheric distillation tower, the main component of a refinery, is where crude oil is heated, evaporated, and then condensed into fractions according to their boiling points. Gases, naphtha, kerosene, diesel, and heavier wastes are some of these fractions.

2. Secondary Processing Units: These units are used to further purify and upgrade these fractions after distillation. Reforming, hydrocracking, and catalytic cracking are a few examples of procedures that raise the output and quality of valuable goods. A variety of refined goods, such as jet fuel, diesel, gasoline, and lubricants, are the ultimate product.

DISTRIBUTION: LINKING END USERS AND REFINERIES

1. Pipeline Networks: These fuels are transported to distribution centers and end customers by a vast network of pipelines, which serve as a lifeline once the refined goods are ready.

Refineries are connected to major cities, airports, and industrial hubs by pipelines that transcend continents. These networks' effectiveness guarantees a consistent and dependable flow of energy resources to satisfy the needs of various markets.

2. Logistics and Storage Facilities: A complex network of logistics and storage facilities is essential to the downstream industry. Large amounts of refined goods are kept in storage on tank farms, which serve as vital reservoirs to keep supply and demand in check.

Vehicles such as trucks, trains, and ships function as mobile conduits, carrying petroleum to locations where pipelines are not accessible or when adaptability is necessary to accommodate fluctuating demand trends.

BEYOND FUELS IN MANUFACTURING AND PETROCHEMICALS

1. Production of Petrochemicals: The downstream oil sector goes well beyond just producing fuels; it also produces petrochemicals. Numerous goods are made from building blocks extracted from crude oil, such as benzene, propylene, and ethylene. Petrochemical manufacturing is a sophisticated process that produces a wide range of daily commodities, including plastics, synthetic rubber, and solvents.

2. Manufacturing Industries: The foundation of many manufacturing industries is downstream products. Refined fuels are necessary for the automotive industry, and petrochemicals are used to make medicines, textiles, and plastics. Thus, the downstream oil sector drives industrialization and technical growth by serving as a vital enabler for a variety of economic activities.

INNOVATION AND ENVIRONMENTAL ASPECTS

1. Environmental Impact: Because of its effects on the environment, the downstream oil sector is coming under more and more scrutiny. Both the burning of fossil fuels and the emissions from refining operations are factors in air pollution and global warming. The sector is facing pressure to reduce its environmental impact by implementing cleaner technology and investigating alternative energy sources.

2. Technological Developments: Innovation in the downstream industry encompasses more than just refining procedures; it also includes operational effectiveness and environmental stewardship. Research and application into advanced technologies, like carbon capture and utilization, cleaner refining techniques, and biofuel creation, are ongoing.

These companies manage the distribution and sale to end users also. Some downstream companies of a few countries are listed below:

INDIA

- Indian Oil Corporation Limited (IOCL)
- Bharat Petroleum Limited (BPCL)
- Hindustan Petroleum Limited (HPCL)
- Bongaigaon Refineries
- Mangalore Refineries Ltd
- Cochin Refinery (Kochi Refinery)
- Assam Oil Division
- Numaligarh Refineries Ltd

- ➤ Petronet LNG Limited
- ➤ Reliance Industries Limited
- ➤ Nayara Energy (Earlier Essar Oil)

FRANCE
- ➤ Engie
- ➤ Schlumberger

CHINA
- ➤ China National Petroleum Corporation (CNPC)
- ➤ Sinopec Shanghai Petrochemical Company Limited
- ➤ Shell Energy (China) Limited
- ➤ Chevron Corporation
- ➤ Total SA

CANADA
- ➤ Imperial Oil Limited
- ➤ Suncor Energy Inc.
- ➤ Royal Dutch Shell Plc.
- ➤ Husky Energy Inc.
- ➤ Irving Oil Ltd

GERMANY: Wintershall Dea

AFRICA
- ➤ Nigerian National Petroleum Company Limited
- ➤ Egyptian General Petroleum Corporation
- ➤ Shell Plc.

- ExxonMobil Corporation
- Midoil Refining & Petrochemicals Company Limited

ARGENTINA

- Axion Energy
- Pan American Energy SL
- Pampa Energia SA
- Yacimientos Petroliferos Fiscales SA
- Raizen SA

PERU

- Repsol SA
- Total SA
- Royal Dutch Shell Plc.
- Petroleos del Peru SA
- CF Industries Holdings Inc.

USA

- ExxonMobil
- Chevron
- Total
- Shell
- BP
- Eni

AUSTRALIA
- Santos
- Woodside Energy
- Origin Energy

RUSSIA
- Gazprom
- Lukoil
- Novatek
- Rosneft
- Surgutneftegas
- Tatneft
- Transneft

SAUDI ARABIA: Saudi Aramco

QATAR: Qatar Petroleum

OMAN: Petroleum Development

IRAQ: Iraq National Oil Company

IRAN: National Iranian Oil Company

KUWAIT: Kuwait Petroleum Corporation

INDONESIA: Pertamina

JAPAN
- Cosmo Oil Company
- Idemitsu Kosan
- Inpex
- Technip FMC
- Total Energies SE

Chapter Two

CRUDE OIL EXPLORATION

In this introductory chapter, we will explore the geological formation of crude oil and the variables that contribute to the construction of large subsurface reservoirs. This chapter lays the groundwork for a deep dive into the world of crude oil, covering everything from the principles of petroleum geology to the exploration technologies that reveal the hidden treasures beneath the Earth's surface.

Edwin L. Drake is universally referred to as "The Founder of the Petroleum Industry". The first production of Crude oil came from Drake Oilfield in Pennsylvania, USA in 1859, which produced only 30 barrels per day from this well.

Before 1869 Russia's oil wells were dug by hand, but in the same year Rudolph and Ludwig Nobel introduced deep drilling at Baku after which the output increased drastically.

The crude production increased with time as more and more countries adopted modern methods, and in the year 2021 the world production of crude oil reached an average of 74 million barrels per day. Saudi Arabia is the largest crude oil exporter followed by Russia.

Exploration for crude oil is the first chapter in the fascinating story of the petroleum industry; it reveals the liquid gold reserves hidden beneath the surface of the Earth that power our contemporary civilization. In order to locate the hidden treasures that might power nations and satisfy the world's unquenchable need for energy, explorers must navigate a complex and scientifically driven process that combines geology, technology, and creativity.

The knowledge of Earth's geological formations is fundamental to the exploration of crude oil. In this stage, geologists are essential. They use a variety of instruments and methods to interpret the language etched in the rocks below the surface of the Earth. First, a thorough examination of the surface geology is conducted, looking at rock formations and outcrops that may provide evidence of the presence of hydrocarbons beneath the surface.

SEISMIC SURVEYS: REVEALING HIDDEN TREASURES

The seismic survey is a fundamental method used in contemporary crude oil exploration that enables geoscientists to see beneath the surface of the Earth and reveal its secrets. Seismic waves, which are sound waves that are generated under control at the surface and bounce back when they come into contact with different rock layers, are the process at work here. Subsurface structures that could potentially contain oil and gas reservoirs are revealed by analyzing the waves that return. This process produces comprehensive photographs of the subsurface.

Remote Sensing and Satellite Technology:

Crude oil explorers use satellite technology and remote sensing to increase efficiency and obtain complete data. Identification of surface characteristics, anomalies, and possible markers of subsurface hydrocarbons is made easier with the use of high-resolution satellite photography. This aerial perspective improves the accuracy of the exploratory work, enabling a more focused and knowledgeable strategy.

Investigative Drilling: The Crucial Moment

Although remote sensing and seismic surveys offer priceless insights, exploratory drilling offers the most definitive validation. In order to verify the existence of oil or gas, wildcat wells are dug at strategic locations determined by the interpretation of geological data. Geologists may see into the Earth's subsurface with the help of cores taken from these wells, which provide vital details on the sorts of rocks, their porosity, and possible hydrocarbon reserves.

Cutting-Edge Well Logging and Drilling: While Logging: Cutting-edge well logging technologies are utilized as drilling advances. To assess the physical characteristics of the rocks, such as resistivity, porosity, and permeability, logging tools are dropped into the well. Real-time data collection during logging allows for quick decision-making and drilling plan modifications in response to encountered subsurface conditions.

Assessing Reservoir Capacity:

Assessing reservoir potential involves more than just verifying the presence of hydrocarbons. The size, quality, and producibility of the reservoir are evaluated by geoscientists and reservoir engineers. Pressure, temperature, and fluid composition are among the parameters that are carefully examined in order to assess the discovery's commercial feasibility.

Environmental Factors to Be Considered

There are difficulties involved in exploring for crude oil, and environmental issues must be taken very seriously. Extensive environmental impact studies are carried out in order to reduce the environmental impact of exploratory activities, safeguard ecosystems, and comply with strict legal requirements. Responsible exploration efforts must incorporate sustainable techniques like wildlife conservation and reduced-impact drilling.

Origin of Oil & Gas

Oil and Natural gas are made up of hydrocarbons (chains of carbon and hydrogen atoms), which are naturally found in rocks inside earth's crust Crude.

Hydrocarbons are produced by a naturally occurring organic soluble material called "KEROGEN" which is found in source rock.

There are 3 types of Kerogen:-
   a) Marine algae and plankton produce liquid oil
   b) Marine and terrestrial plant and animals produce oil and/or gas
   c) Terrestrial plants and animals produce Coal

Technically, when buried deep inside the earth, over time, these Kerogen-laden rocks are cracked with heat into smaller hydrocarbon molecules making liquid oil and gas.

This oil/gas comes in contact with rocks with enough porosity and permeability and pass through it up to an oil well.

The formation of crude oil/gas stepwise is as under:

a) The compression of plant an animal remains inside the earth gets decayed over time in sedimentary rocks such as limestone and shale.

b) As the layers of sediments get deposited on the ocean floor, the decaying remains of animals and plants get deposited on these forming rocks.

c) The organic material so formed ultimately transforms into oil and gas after being exposed to specific temperature and pressure within the earth's crust.

Today when a new oil well or gas well is explored, one of the early duties is to have a chemist test the gas for its "gasoline content".

Many refineries today own chains of distributing companies & retail network, thereby safeguarding the market for their products as well as the profits.

The story of crude oil's origins unfolds deep into our planet's geological records. This hydrocarbon-rich liquid that powers our modern world has a complex history molded by geological processes that have lasted millions of years. Understanding the geological origin of crude oil necessitates diving into the complex interactions between organic matter, sedimentary environments, and the transformational forces under the Earth's surface.

Hydrocarbons' Organic Origins

Crude oil is formed by the leftovers of small marine organisms that once thrived in ancient oceans. These creatures, principally plankton and algae, were crucial in the formation of crude oil's organic precursor. When these species died, their organic remains sank to the ocean floor, generating layers of organic-rich sediments.

The organic matter in these layers was gradually changed into a waxy substance called Kerogen through a process called diagenesis, which was triggered by a combination of warmth and pressure. Crude oil and natural gas originate from Kerogen. The source material and the particular geological circumstances during diagenesis determine the kind and makeup of the Kerogen.

Maturation and Burial: Geological Change

The buried organic material changed even more when sediment layers built up on top of the organic-rich deposits due to the rising pressure from the surrounding sediments. By exposing the Kerogen to increased temperatures and pressures, a process called as maturation or thermal cracking, it eventually transformed into liquid and gaseous hydrocarbons.

It is crucial to know the depth at which this maturation takes place since it establishes whether the hydrocarbons will mostly become natural gas or crude oil. Crude oil typically originates at shallower depths, but natural gas is produced deeper in the Earth at higher temperatures and pressures.

Migration: The Subterranean Journey

After it is produced, crude oil travels through the complex network of pore spaces in rocks on an underground voyage. Faults and folds in the geology, as well as the permeability of the rocks that oil passes through, affect its movement. The oil may travel great distances from the rocks that are its source to reservoirs where it may gather.

Formation of Reservoirs: Capturing the Liquid Gold

Crude oil must be trapped in geological formations large enough to store hydrocarbon volumes in order for it to become a viable resource. Crude oil is stored in reservoir rocks, which are usually permeable and porous sandstones or carbonate rocks. These rocks' permeability makes it easier for oil to travel inside the reservoir, while their porosity permits the storage of liquids.

These reservoirs contain a variety of trapping techniques that keep oil contained. Oil naturally collects in natural traps behind impermeable layers created by the bending of rock layers. Faults, or fissures in the crust of the Earth, can serve as barriers, stopping the migration of oil and causing isolated accumulations. Another mechanism for capturing oil is the formation of stratigraphic traps, which are caused by differences in sedimentary rock layers.

Discovering and Exploration: Uncovering the Secrets of Earth

Understanding the geological hints buried beneath the surface of the Earth is essential to the search for crude oil. Geologists use a range of exploration methods to find possible hydrocarbon resources. For instance, seismic surveys employ sound waves to provide precise images of subsurface structures, which help geologists, locate possible reservoirs and traps.

Finding the geological formations that contain crude oil also requires drilling exploratory wells. These wells yield cores, which are a rich source of information on the composition, composition of the rock layers, and presence or absence of hydrocarbons. The physical qualities of rocks are measured with the use of logging instruments, which helps determine the characteristics of the reservoir.

Different Types of Crude Oil Reservoirs

There are numerous varieties of crude oil reservoirs, each with specific geological traits. For effective extraction and production, one must have a thorough understanding of various reservoir types.

1. Conventional Reservoirs:

Permeable rocks that facilitate easy oil flow, such sandstone or limestone, are what define conventional reservoirs. These reservoirs usually have distinct traps, and conventional drilling techniques can be used to retrieve the oil.

2. Unconventional Reservoirs:

Greater geological difficulties arise from unconventional reservoirs. Because shale rocks, for example, have low permeability, stored oil must be released using sophisticated extraction methods like hydraulic fracturing, or fracking. One well-known example of an unconventional oil reservoir is the Bakken Formation in North Dakota and Montana.

3. Reservoirs of Heavy Oil:

Heavy oil reservoirs contain oil with a higher viscosity, which makes extraction more difficult. To boost the flow of oil, these reservoirs frequently require enhanced oil recovery (EOR) procedures like as steam injection. The Athabasca Oil Sands in Canada are famous for their heavy oil concentrations.

TYPES OF CRUDE OIL

Crude oil mainly consists of hydrocarbons and varies in appearance and consistency from country to country. Each crude oil is unique and range from yellow-brown liquid to black viscous semi solids. Petroleum found in nature is a liquid lighter than water, normally opaque but of a dark brown colour in thin layers by transmitted light, and often dark green by reflected light.

The composition of crude oil is a complex mixture of hydrocarbons that varies greatly depending on the location and geological characteristics of the oil reservoir.

The smell of Italian light-coloured oil is pleasant and has a distinctive aromatic odor. There are other oils in which the smell is most unpleasant. This is frequently coincident with the presence of Sulfur and nitrogen compounds.

The main constituents of Petroleum are a large number of compounds of carbon and hydrogen combined in various proportions. Occasionally, compounds of carbon, hydrogen and nitrogen, or carbon, hydrogen and oxygen, or carbon, hydrogen and Sulfur, are the main constituents of petroleum.

The hydrocarbons may be with simple 1 carbon atom molecule (methane) or with 50 or more carbon atoms; the main property of any hydrocarbon depends solely on the number of carbon atoms in the molecule. Sometimes hydrogen sulphide is also contained in the solution (hydrocarbon mix).

Different crude oils contain varying amounts of:

- Paraffins
- Aromatics
- Naphthenes
- Lower hydrocarbons
- Higher hydrocarbons
- Dissolved gases

Olefins are generally not present in the hydrocarbons, but are produced in crude processing operations involving gasoline production.

The nature of products to be obtained depends mainly on the nature of crude processed. Crude oil is generally classified into 3 categories according to the nature of the hydrocarbons:

1. **Paraffin base crude oils:** These contain paraffin wax and very little or no asphaltic matter. They give good yield of paraffin wax and high-grade lubricating oils.
2. **Asphaltic base crude oils:** These contain little or no paraffin wax but the asphaltic matter is usually present in large quantities. They consist mainly of Naphthenes and give good yield of lubricating oils. These crudes are also referred as Naphthenes base crude oils.
3. **Mixed base crude oils:** These crudes contain substantial amounts of both paraffin wax and asphaltic matter.

Crude oil is divided into different categories according to its physical attributes, density, and sulfur content. Here are a few popular varieties of crude oil:

1. Brent Crude oil:

Origin: The Brent oil field in the North Sea.

Features: Low sulfur concentration and light, sweet crude. About two-thirds of the world's traded crude oil is priced using Brent crude, which is a significant benchmark for oil prices worldwide.

2. West Texas Intermediate (WTI):

Origin: mostly from Texas fields, in the United States.

Features: Light, sweet crude with a reduced sulfur concentration. WTI is a significant additional oil price benchmark that is used to determine the price of oil produced in the Americas.

3. Dubai Crude:

Origin: Middle East, notably from United Arab Emirates fields.

Features: Sulfur content is higher in this medium-sour crude than it is in WTI and Brent. The benchmark for oil prices in the Asian market is Dubai crude.

4. OPEC Basket:

Origin: Member nations of the Organization of the Petroleum Exporting Countries (OPEC).

Features: This is a weighted average of oil prices from OPEC member nations rather than a particular kind of crude oil. It consists of a blend of different crude oils produced by OPEC members.

5. Bonny Light

Origin: Nigeria, specifically the Niger Delta.

Features: Low sulfur level, light and sweet crude. Bonny Light is a highly esteemed crude oil recognized for its superior qualities.

6. Urals Blend:

Origin: Russia, mostly in the area of the Ural Mountains.

Features: Mildly acidic crude. A prominent Russian export grade, Urals Blend is frequently used as a reference price for Russian crude oil.

7. Mars Blend:

Origin: Gulf of Mexico, United States.

Features: Mildly acidic crude. In the Gulf of Mexico, Mars Blend is a significant crude grade that is used as a benchmark for medium-sour crude prices in the US market.

8. Tapis Crude:

Origin: Malaysia.

Features: Sweet, light crude, in the Asia-Pacific area, Tapis is a vital crude oil that is frequently used as a benchmark for pricing crude oil in this market.

9. Maya Crude:

Origin: Mexico.

Features: Strong, acidic crude, Maya is a notable Mexican crude oil grade distinguished from lighter crudes by its increased density and sulfur concentration.

It's crucial to remember that there are numerous additional grades of crude oil in the world; these are only a few examples. The differences between different kinds of crude oils affect their market value, the amount of refining that is necessary, and the products that are made from them.

Furthermore, the division of products into "light" or "heavy" and "sweet" or "sour" categories aids in conveying important features to industry players.

CHARACTERISTICS OF CRUDE OIL

The two main properties for classifying crude oil are API Gravity and Sulphur content.

API GRAVITY:

In the petroleum industry, the density of crude oil is expressed in °API (degrees API). API gravity varies inversely with the density, i.e., lighter material has higher API value.

The API gravity of crude is a measure of how heavy or light the crude oil is, in simple terms. Lighter crudes with API above 38 degrees give better yield of gasoline, diesel and jet fuel (all these are in high demand). Heavy crudes give more yield of asphalt, furnace oil, heavy furnace oil and other heavy petroleum products.

SULFUR CONTENT:

Amongst all the elements present in the crude oil, sulfur has the most profound effect in the refining process. Higher levels of sulfur can cause:

- Corrosion of refining equipment
- Deactivate the catalysts used in refining process
- Lead to undesirable air emissions of sulfur compounds which are subject to strict regulatory controls

Presence of sulfur in vehicle fuels (Gasoline, diesel) leads to undesirable emission of organic compounds of sulfur which is a health hazard and subject to strict government regulations in most countries.

Refineries have to spend big amounts in maintaining their capability to manage Sulphur emissions. In common terms the crude is also called as sweet crude (sulfur content less than 0.5% by weight) or sour crude (sulfur content 1.0 to 2% by weight, in some cases it goes up to 4% also).

The sulfur content and API gravity of some of the world's crude oil are as under:

Crude oil	Country of origin	Type of crude oil	Gravity °API	Sulfur %
Arabian light	Saudi	Medium sour	34	1.9
Kuwait	Kuwait	Medium sour	30.9	2.5
Daqing	China	Medium sour	33.0	0.1
Forcados	Nigeria	Medium sour	29.5	0.2
West Texas	USA	Light sweet	39.8	0.3
Brent	UK	Light sweet	40.0	0.5
Maya	Mexico	Heavy sour	21.3	3.4

The pricing of crude oils is highly influenced by its quality. Light sweet crudes demand a premium compared to other sour variety of crudes.

Light sweet crudes require less energy to process, and need lower capital investment to meet the desired quality standards of the market.

The price differential between light sweet and sour crude is generally in the range of 15 to 25% (being higher for sweet crude).

The quality of light-sweet and heavy-sour crude varies from place to place and the supply demand scenario plays an equally important role in determining the crude pricing.

Chapter 3

GEOPHYSICS AND SEISMIC TECHNIQUES

Exploration for crude oil is a dynamic and intricate process that mostly depends on seismic and geophysical methods to solve the riddles buried beneath the surface of the Earth. As the industry's eyes and ears below the surface, geophysics offers priceless insights into the subsurface features and structures that might contain hydrocarbon resources. Seismic techniques are a cornerstone among the variety of geophysical technologies available, providing a comprehensive and three-dimensional image of the Earth's interior. We explore the fundamentals, methods, and uses of seismic and geophysics in this thorough investigation, highlighting the vital role these fields play in the search for and extraction of crude oil.

FOUNDATION OF GEOPHYSICS

Recognizing Earth's Subsurface: The discipline of geophysics is based on the knowledge that differences in the physical characteristics of rocks and liquids beneath the surface of the planet can be found and analyzed to reveal subsurface geological features and possible hydrocarbon reserves. Seismic velocity, electrical conductivity, magnetic susceptibility, and density are some of these physical characteristics.

Geophysics methods Overview: There are a number of different techniques that fall under the broad category of geophysical methods, each with a specific function in the exploration process. Seismic, gravity, magnetic, electrical resistivity, and well logging surveys are a few of the important geophysical techniques. In the oil and gas sector, seismic surveys in particular are now the main method for mapping underlying structures.

SEISMIC TECHNIQUES IN CRUDE OIL EXPLORATION

SEISMIC METHODS FOR EXPLORATION OF CRUDE OIL

1. Fundamentals of Seismic Investigation:

In order to measure the reflections and refractions of seismic waves as they pass through subsurface layers, seismic techniques generate acoustic waves (also known as seismic waves) at the Earth's surface. The underlying idea is that varied rock formations and types have varying seismic velocities, which makes it possible to create intricate representations of the subsurface.

2. Components of a Seismic Survey:

Source: A controlled energy source, such as an air pistol in maritime surveys, explosives, or a vibrator truck, creates seismic waves in a seismic survey that travel through the Earth.

Receivers (Geophones or Hydrophones): These sensors record the reflected and refracted seismic waves and are strategically positioned on the sea floor or the Earth's surface.

Imaging and Processing: To provide detailed photographs of subsurface structures, the captured data is processed using advanced signal processing techniques. Algorithms are used in this process to improve signal quality, reduce noise, and produce visual depictions of the subsurface.

3. Types of Seismic Surveys:

2D Seismic surveys: Two-dimensional profiles of the subsurface are provided by 2D seismic surveys, which are frequently the initial stage of a new area's exploration.

3D Seismic Surveys: These more sophisticated and extensively utilized surveys offer a three-dimensional picture of the subsurface. Geological structures may now be represented with greater accuracy and detail thanks to technological advancements.

4. Reflection Seismology: Reflection Coefficient: Part of the energy from seismic waves is reflected back to the surface when they come into contact with a border between two distinct rock layers. An essential piece of information regarding the subsurface composition is provided by the reflection coefficient, which the ratio is of reflected to incident energy.

5. Refraction Seismology: Refraction Coefficient: In refraction seismology, waves arrive at receivers with a temporal delay due to their distinct velocities as they pass through subsurface strata. The depth and velocity of subsurface layers can be estimated with the aid of the refraction coefficient.

APPLICATIONS OF SEISMIC TECHNIQUES IN EXPLORATION

1. Mapping Subsurface Structures: Seismic techniques play a critical role in mapping subsurface structures, assisting geologists in identifying possible hydrocarbon reserve locations. Exploration teams can make informed decisions on where to drill exploratory wells when they can visualize geological formations.

2. Determining Rock Properties: Seismic data is used to understand the properties of rocks within buildings as well as to image them. Different rock types have different seismic velocities, which aids in the identification of reservoir rocks, seals, and probable hydrocarbon traps.

3. Reservoir characterization: Seismic techniques contribute to reservoir characterization in addition to detection. This includes determining the porosity, permeability, and fluid content of the rocks. This information is critical for calculating a reservoir's potential productivity.

4. Hydrocarbon Type and Saturation: Seismic data can also reveal the type of hydrocarbons present (oil, gas, or a combination) and where they are distributed within the reservoir. Understanding saturation levels aids in the optimization of manufacturing processes.

5. Monitoring Reservoir Changes: Seismic monitoring techniques are used during and after the production phase to track reservoir changes. This enhances reservoir management by allowing operators to alter production tactics and improve recovery.

ADVANCED SEISMIC TECHNOLOGIES

1. 4D Seismic Surveys: Time-Lapse Imaging: Repetitive seismic observations over the same region at various times are part of 4D seismic surveys. With the help of time-lapse photography, it is possible to track how the reservoir changes over time and gain insight into how industrial activities affect it.

2. Shear-Wave Imaging in Multicomponent Seismic Surveys: While conventional seismic surveys monitor compressional waves, multicomponent surveys also record shear waves. The characterization of reservoirs is improved by shear-wave imaging, especially in unconventional plays where shear-wave information is useful.

3. Full Waveform Inversion (FWI): High-Resolution Imaging: FWI is a sophisticated inversion method that compares modeled and observed seismic data iteratively in order to improve subsurface models. High-resolution imaging is made possible by this technique, which offers comprehensive details about intricate underlying structures.

4. Seismic While Drilling (SWD): Drilling wells while gathering seismic data; this allows for real-time decision making. Drilling plans can be modified instantly based on new insights into subsurface conditions thanks to this real-time data.

CHALLENGES AND INNOVATIONS IN SEISMIC EXPLORATION

Challenges: Complex Geological Settings

Accurately understanding subsurface structures is a challenge when conducting seismic research in regions with complicated geological settings, such as salt domes or unconventional reservoirs.

Environmental Impact:

The use of explosives in seismic surveys, especially in maritime regions, may have an impact on the environment. As a result, attempts are being made to create seismic sources that are quieter and more environmentally friendly.

2. Innovations:

AI and Machine Learning: Workflows for seismic interpretation are increasingly using AI, machine learning, and advanced data analytics. The speed and precision of subsurface imaging and interpretation are improved by these technologies.

Quantum Seismic Sensors:

With the development of quantum sensor technology, seismic measurements could be revolutionized. The quality of seismic data could be enhanced by these sensors' capacity to produce measurements that are more sensitive and accurate.

Geophysics and seismic techniques are key players in the complex dance of crude oil discovery, providing hitherto unattainable windows into the Earth's subsurface.

Chapter 4

DRILLING TECHNOLOGIES AND WELL CONSTRUCTION

Modern drilling methods are essential to the intricate and sophisticated process of extracting crude oil from beneath the surface of the Earth. The oil and gas sector has had significant breakthroughs in drilling techniques, instruments, and apparatus over time, which have transformed the effectiveness and security of exploration and production endeavors. This thorough investigation explores the wide range of crude oil drilling technology, from conventional drilling techniques to the most recent advancements influencing the direction of the sector.

In United States 2 types of drilling methods are used for different rock and soil conditions:

1. Standard cable tool system (Percussion system)
2. Rotary-hydraulic system

Cable Tool Drilling (Percussion system)

Figure 3: Standard Cable Tool Drilling System

One of the first techniques employed in the oil industry was cable tool drilling, sometimes referred to as percussion drilling and it is so called because the process employs repeated hammering. Using this technique, a heavy string of drilling instruments is raised and lowered repeatedly in order to crush and cut through the rock. Although it is now mostly unnecessary for extracting oil, it was vital to the early development of the sector.

In standard Cable-tool drilling, a hole is made by the repeated blows generated by lifting and dropping a heavy chisel bit on rocks or underground. The heavy chisel bit works as the "tool" and the cable could be made of a manila rope or multiple steel strands.

Percussion drilling, utilizing impact forces to break rock, was an early technique that evolved into cable tool drilling. While it has been largely replaced by rotary drilling, percussion methods are still used in specialized situations, such as mining and certain geotechnical applications.

Rotary-hydraulic system

When rotary drilling was developed in the late 1800s, it completely changed the drilling industry. Drilling mud is circulated to cool the bit, raise cuttings to the surface, and maintain pressure as a drill bit is rotated to produce a borehole. Modern drilling operations are based on rotary drilling, which has continuously improved in terms of efficiency and depth capabilities.

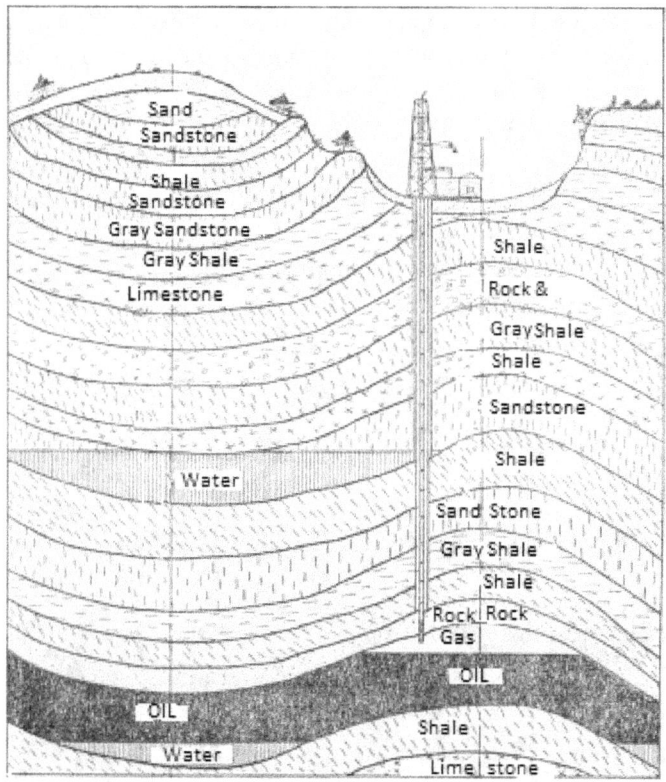

Figure 4 : Drilling layers of rock & soil

The Rotary-hydraulic system is mainly used in the gulf coast fields where rocks lie above the oil-bearing sands. It is similar to a machinist drilling a hole through a casting. The drill stem is hollow and under pressure of 40 to 100 psi, a mixture of mud and water is forced through which forces the fluid to escape through the drill bit at the bottom of the well. Powerful pumps keep this fluid in constant circulation.

Once started the drilling usually goes on day and night until the well is completed. The operation is done in 12 hour shifts also called as "tours". The crew usually consists of 8 men, 4 on each tour, a driller, a tool dresser, a helper and an engineer.

The drillers are very well conversant with the system and the driller can tell what is going on at the bottom of the hole 200 feet deep or even 4000 feet deep. Many unfortunate and costly things however do happen during the drilling of a well - the cable may break resulting in finishing the job that may last days, weeks or even months, it may even cause the abandonment of the well. One man, usually called the "pumper," can take care of a number of wells.

The cost of wells varies with their depth, the number, hardness and thickness of rock strata, number, size and length of strings of casing, fluctuations in the cost of material and labor, distance from shipping point, etc.

CONVENTIONAL ROTARY DRILLING

COMPONENTS OF ROTARY DRILLING SYSTEM:

Drill Bit: A cutting instrument used to smash and pierce rock.

Drill Pipe: A network of pipes that transmits rotating motion and drilling mud to the bit.

Rotary Table: The surface component that turns the drill string is known as a rotary table.

Mud Pumps: circulate drilling mud, a fluid used to cool, lubricate, and transport cuttings to the surface.

Derrick: A framework that holds the drill string and other drilling components in place.

TYPES OF ROTARY BITS:

Roller Cone Bits: These bits have rotating cones with teeth that shatter the rock.

Fixed Cutter Bits: For increased durability, use diamond or other cutting elements.

Polycrystalline Diamond Compact (PDC) Bits: Diamond-enhanced cutters are used to improve efficiency.

DRILLING FLUIDS:

Drilling fluids, often known as mud, fulfill several functions in rotary drilling. They provide buoyancy to the drill string, cool and lubricate the drill bit, and transport cuttings to the surface for analysis. Depending on the geological conditions, various drilling muds are used.

DIRECTIONAL DRILLING:

Directional drilling allows operators to stray from vertical paths, allowing them to access deposits beneath obstacles or in difficult terrain. This method has proved critical in increasing reservoir recovery.

ADVANCED DRILLING TECHNOLOGIES

HORIZONTAL DRILLING:

Horizontal drilling entails turning the drill bit horizontally once it has reached a specific depth. This approach allows for more contact with the reservoir, which improves production rates and overall recovery.

EXTENDED REACH DRILLING (ERD):

ERD techniques allow drilling to access reserves located far away from the drilling site. This method reduces the number of drilling locations required to access large reservoirs, decreasing the environmental impact.

UNDERBALANCED DRILLING (UBD):

Underbalanced drilling includes keeping the pressure in the wellbore lower than the formation pressure. This approach reduces reservoir rock damage while increasing hydrocarbon recovery efficiency.

COILED TUBING DRILLING (CTD):

Coiled tubing drilling transports drilling mud and deploys down hole tools via a continuous, flexible tube. This approach is effective in some applications, resulting in cost savings and a lower environmental impact.

OFFSHORE DRILLING TECHNOLOGIES

JACK-UP RIGS:

Jack-up rigs are transportable drilling platforms that sit on seabed legs. These rigs are typically utilized in shallow water and can be hauled to different areas.

SUBMERSIBLE RIGS:

Submersible rigs can float to a drilling spot before submerging to the seafloor. In comparison to jack-up rigs, this technology enables for drilling in deeper waters.

DRILLSHIPS:

Drill ships are vessels that are outfitted with drilling equipment and capabilities. They are useful for offshore exploration since they are mobile and can drill in deep water.

SEMI-SUBMERSIBLE RIGS:

These are floating platforms with submerged pontoons. They are excellent for offshore drilling because they provide stability and can work in deep-water settings.

DRILLING CHALLENGES

1. HIGH-PRESSURE, HIGH-TEMPERATURE (HPHT) ENVIRONMENTS:

These conditions include high pressure and temperature. To overcome these obstacles, advanced materials and drilling fluid technologies are used.

2. UNCONVENTIONAL RESERVOIR DRILLING:

Drilling methods specific to unconventional reservoirs, such as shale formations, is needed. Fracking, also known as hydraulic fracturing and horizontal drilling are popular techniques for removing hydrocarbons from these reservoirs.

3 WELL LOGGING AND CORING:

By removing cylindrical rock samples for in-depth examination, coring offers information into the properties of the reservoir. Many instruments are used in well logging to measure parameters such fluid saturation, porosity, and resistivity.

4. DRILLING WASTE MANAGEMENT:

A crucial part of drilling operations is the management of drilling waste, which includes cuttings and mud. Environmental stewardship requires sustainable activities like reusing drilling fluids and disposing of debris properly.

INNOVATIONS AND FUTURE TRENDS

AUTONOMOUS DRILLING SYSTEMS:

Autonomous drilling systems are becoming a reality as automation and artificial intelligence become more integrated. These devices improve safety, minimize the need for human intervention, and optimize drilling operations.

DRONES AND ROBOTICS:

Drones and robotics are being used for duties including drilling equipment monitoring, maintenance, and inspection. The effectiveness and safety of drilling operations are increased by these technologies.

DIGITAL TWIN TECHNOLOGY:

This technique enables real-time monitoring and analysis by building virtual copies of physical assets. Digital twins in drilling improve performance, forecast equipment breakdowns, and facilitate better decision-making.

GREEN DRILLING TECHNOLOGIES:

The development of ecologically friendly drilling technologies is a growing industry priority. To lessen the environmental impact of drilling operations, this involves using sustainable techniques, cutting emissions, and implementing alternative energy sources.

Crude oil drilling techniques cover a wide range and are always changing. The industry has advanced significantly from the days of cable tools and rotary drilling to the complex autonomous systems and digital twin technologies of today.

Crude oil exploration and production are shaped by innovation and the ongoing pursuit of ecologically conscious, safe, and efficient drilling techniques. The combination of conventional knowledge and state-of-the-art technologies accelerates the sector toward new frontiers as we negotiate the depths of extraction, guaranteeing a resilient and sustainable energy future.

Chapter – Five

RESERVOIR ENGINEERING AND MANAGEMENT

A vital component of the oil and gas sector, reservoir engineering and management is the search, development, and extraction of hydrocarbons from subterranean reservoirs. In order to maximize the recovery of oil and gas resources, this multidisciplinary field integrates concepts from geology, fluid mechanics, mathematics, and other technical disciplines. This in-depth examination of reservoir engineering and management will cover the major ideas, approaches, difficulties, and innovations influencing this ever-evolving area.

A specialist area of petroleum engineering called reservoir engineering is concerned with comprehending how hydrocarbons behave in subterranean reservoirs. The main objective is to provide effective reservoir management while optimizing the economic recovery of gas and oil. Reservoir characterization, fluid flow analysis, well testing, and the creation and improvement of production plans are important aspects of reservoir engineering. . The key features being:-

1. **Reservoir Characterization:**

    The first step in the process involves the collaboration of reservoir engineers and geoscientists to collect and examine subsurface data. This contains details about the geometry of the reservoir, the fluid and rock characteristics, and other elements that are essential to comprehending the behavior of the reservoir

2. **Fluid Flow Analysis:**

   A key component of reservoir engineering knows how fluids flow within a reservoir. Modeling the behavior of water, gas, and oil in response to variations in temperature, pressure, and rock characteristics is known as fluid flow analysis. A number of mathematical models are used to simulate fluid flow in reservoirs, including the material balance equation and Darcy's law.

3. **Well Testing:**

   One of the most important methods for evaluating reservoir performance is well testing. It entails gathering information from individual wells in order to assess the permeability and skin factor of the reservoir. In order to understand reservoir behavior and interpret well test data, pressure transient analysis is frequently employed.

4. **Reservoir Simulation:**

   To simulate a reservoir, numerical models of the subsurface reservoir's fluid dynamics must be made. To model reservoir conditions, forecast production behavior, and enhance recovery tactics, sophisticated software tools are utilized. When it comes to well placement, injection tactics, and production rates, simulation models help engineers make well-informed judgments.

KEY CONCEPTS IN RESERVOIR ENGINEERING

a) Mechanisms for Driving Reservoirs:

A variety of mechanisms propel hydrocarbons into the wellbore from reservoirs. Natural depletion, water flooding, gas injection, and thermal techniques are the main drive mechanisms. Choosing the best recovery plan requires an understanding of these systems.

b) Recovery Factor:

The percentage of hydrocarbons that may be profitably extracted from a reservoir is indicated by the recovery factor. It is affected by fluid parameters, reservoir features, and the efficiency of recovery methods. In reservoir engineering, increasing the recovery factor is a primary objective.

c) Enhanced Oil Recovery (EOR):

Enhanced Oil Recovery, or EOR, is the process of extracting more oil from a reservoir by using procedures that go beyond conventional recovery methods. Water flooding, gas injection (such as CO_2 or natural gas), and thermal techniques like steam injection are common EOR techniques. In order to improve hydrocarbon recovery, these methods try to change the physical or chemical characteristics of the reservoir.

d) Management of Reservoirs:

Achieving optimal recovery from reservoirs requires making decisions that take into account social, environmental, and economic aspects. This entails keeping an eye on reservoir performance, putting production plans into action, and gradually adjusting to shifting reservoir circumstances.

CHALLENGES IN RESERVOIR ENGINEERING

Risk and Uncertainty:

Because reservoirs are intricate systems, there are always going to be unknowns. Accurately predicting reservoir behavior is difficult due to the dynamic nature of subsurface conditions and the variability in rock and fluid parameters. To make wise judgments, reservoir engineers must manage uncertainty and evaluate risks.

Reservoir Heterogeneity:

Fluid flow patterns are impacted by the heterogeneity of reservoir rocks, which are distinguished by variances in permeability, porosity, and other characteristics. Reservoir heterogeneity calls for sophisticated modeling methods as well as flexible reservoir management plans.

Environmental Considerations:

Reservoir engineering and management extend beyond technical aspects to address environmental concerns. Minimizing the environmental impact of hydrocarbon extraction, managing produced water, and ensuring the integrity of wellbore constructions are critical considerations.

Economic Factors:

Economic factors play a pivotal role in decision-making. Fluctuations in oil and gas prices, operational costs, and investment requirements influence the feasibility of recovery projects. Reservoir engineers must balance economic considerations with technical goals.

INNOVATIONS IN RESERVOIR ENGINEERING

Advanced Reservoir Imaging Techniques:

New advancements in geophysical technology, like electromagnetic techniques and 3D seismic imaging, offer more precise and in-depth pictures of subterranean reservoirs. These sophisticated imaging methods improve the characterization of reservoirs and help create more accurate reservoir models.

DATA ANALYTICS AND MACHINE LEARNING:

The way that reservoir engineering has integrated data analytics and machine learning has completely changed how big datasets are interpreted. Based on real-time data, these technologies aid in pattern recognition, reservoir behavior prediction, and production strategy optimization.

Nanotechnology for enhanced oil recovery (EOR):

Nanotechnology is showing promise as a field to improve oil recovery. Injecting nanoparticles into reservoirs can change the fluid and rock characteristics, increasing the effectiveness of sweep operations and overall recovery rates. This creative strategy has the ability to improve productivity while lessening its negative effects on the environment.

Technologies for Reservoir Monitoring:

For efficient management, a reservoir's conditions must be continuously monitored. Real-time data capture is made possible by innovations like intelligent well completions, fiber-optic sensors, and down hole monitoring tools. This improves the capacity to modify production plans in response to shifting reservoir dynamics.

The foundation of the oil and gas sector is reservoir engineering and management, which fills the gap between underground reservoirs and effective hydrocarbon recovery. Because this field is multidisciplinary, it needs an integrated approach that takes geological, engineering, economic, and environmental factors into account. In order to ensure sustainable and responsible resource exploitation, reservoir engineering will continue to be shaped by continuous innovations and technological breakthroughs.

Reservoir engineers and managers work in the ever-changing oil and gas sector, pushing the limits of technology and expertise to find the best ways to extract hydrocarbons while meeting the changing needs of the environment and society.

Chapter Six

UPSTREAM OILFIELD FACILITIES

The petroleum industry's upstream sector includes operations pertaining to the discovery, extraction, and drilling of natural gas and crude oil. It is the first phase of the energy supply chain, with the extraction of hydrocarbons from subterranean reserves as its main objective.

Economic Significance: The petroleum industry's upstream operations are the cornerstone of the world's energy supply and economic expansion.

Resource exploration: Upstream operations use exploration to find and assess new sources of gas and oil.

The petroleum industry cannot function without oilfield infrastructure, which are vital to the exploration, extraction, and processing of hydrocarbons. Each of these unique and specialized facilities fulfills a particular role in the overall extraction and processing of oil and gas. The following are a few typical kinds of oilfield facilities:

DRILLING RIGS

A drilling rig is a sophisticated piece of industrial machinery used to bore holes into the Earth's subsurface. These rigs are an important part of the oil and gas business, as well as other industries including geothermal energy development, environmental site inspections, and mineral exploration. Drilling rigs are classified into several categories, each built for a unique use or geological condition.

PARTS OF A DRILLING RIG:

1. Mast:

A mast is a long vertical tower or framework that supports drilling equipment and allows drilling instruments to be raised and lowered into the borehole.

2. Crown Block:

The crown block is a collection of pulleys located at the top of the mast that directs the drilling line as it is lifted or lowered. It is essential to the lifting mechanism.

3. Drilling Line:

The drilling line is a cable or wire rope that extends from the crown block to the drilling instruments and allows the drill string to be raised and lowered.

4. Drill String:

The drill string is a set of interconnected drill pipes that stretches from the surface to the borehole's bottom. The drill bit is linked to the drill string at the end.

5. Top Drive:

A top drive is a motorized device that rotates the drill string and drill bit in modern drilling rigs. This is a different take on the typical Kelly drive.

UPSTREAM OILFIELD FACILITIES 53

6. Kelly Drive:

A Kelly is a long, square or hexagonal pipe used in older rigs to transmit rotating motion from the rotary table to the drill string. Top drives have virtually supplanted it in modern drilling.

7. Rotary Table:

The rotary table is a rotating platform on the rig floor that allows the drill string to rotate. It is propelled by a Kelly or a top drive.

8. Swivel:

The swivel allows the drill string to revolve freely while remaining connected to the drilling mud supply. It's up above the Kelly or top drive.

9. Mud Pumps:

Mud pumps move drilling mud (a mix of water, clay, and different chemicals) down the drill string, transporting cuttings to the surface and cooling the drill bit.

10. Derrick:

The derrick is a tall framework or structure that supports the crown block and keeps the drilling operation stable. It is frequently built of steel and may have a lattice construction.

11. Substructure:

The rig's foundation is the substructure, which supports the derrick and provides a sturdy platform for the complete drilling rig.

12. Drawworks:

The drawworks are a collection of hoisting machinery used to raise and lower the drill string. It is powered by engines and has a braking system to manage the drill string's descent.

13. BOP (Blowout Preventer):

The blowout preventer is a safety device installed at the wellhead to prevent the well from leaking uncontrollably. It is critical for well management and safety.

TYPES OF DRILLING RIGS

Figure 5: Onshore Drilling Rig

1. Onshore Drilling Rigs:

Onshore drilling rigs are located on land and are used for exploration and production activities in places with accessible terrain.

These rigs, sometimes known as "land rigs," are the workhorses of onshore drilling. They are available in a variety of sizes and configurations, and they use a drill bit to penetrate the Earth's surface and reach oil and gas deposits.

a. Derrick:

The derrick is a towering structure on a drilling rig that supports the drill string and enables for the lifting and lowering of equipment during drilling operations.

b. Drill String:

The drill string, which consists of the drill bit, drill collars, and drill pipes, is the main component that penetrates the Earth's strata. The revolving drill bit breaks rocks, resulting in the formation of a borehole for further exploration.

c. Mud Circulation System:

Mud is an important component in drilling operations. It cools the drill bit, transports drill cuttings to the surface, and stabilizes the borehole walls. The mud circulation system maintains a steady and controlled flow.

d. Hoisting System:

The hoisting system, which is in charge of lifting and lowering the drill string, is a collection of winches and pulleys that efficiently handles the heavy equipment required in drilling.

Figure 6: Offshore Drilling Rig

2. Offshore Drilling Rigs:

Offshore drilling rigs, which are located in bodies of water, are classified into several varieties, including jack-up rigs, semi-submersible rigs, drill ships, and submersible rigs.

Offshore drilling rigs are marine structures used to explore the seabed for oil and gas. These rigs, which can be floating or fixed to the ocean floor, use modern drilling equipment to harvest hydrocarbons from beneath the seabed. Offshore drilling is critical to satisfying global energy demands, accessing underwater resources, and promoting economic development. Despite technological and environmental obstacles, offshore drilling rigs play an important role in the offshore energy sector by contributing considerably to the exploration and exploitation of key energy resources.

3. Mobile Drilling Rigs:

Mobile drilling rigs are built to be moved from one area to another. This category includes truck-mounted and trailer-mounted rigs.

4. Fixed Platform Rigs:

Fixed platform rigs are permanently installed structures in shallow waters. They include platforms such as jacketed platforms and compliant towers.

5. Directional Drilling Rigs:

Directional drilling rigs are outfitted with the capability of drilling non-vertical wells, allowing for the exploration and extraction of hydrocarbons from reservoirs in certain directions.

6. Drillship:

A drillship is a type of marine vessel that is outfitted with drilling equipment. It can drill in deep water and is frequently employed for offshore exploration.

7. Coiled Tubing Rigs:

Coiled tubing rigs are built for coiled tubing drilling, which involves using a continuous, flexible tube to transport drilling mud and deploy down hole instruments.

Drilling rigs are at the forefront of oil and gas exploration, allowing crucial energy resources to be extracted from under the Earth's surface. Drilling rigs have evolved to meet the demands of difficult geological conditions and environmental considerations, beginning with the early cable tool and rotary drilling methods and progressing to today's advanced technologies.

As the industry progresses, the convergence of ancient wisdom and cutting-edge technologies accelerates crude oil exploration and production into new frontiers, assuring a sustainable and resilient energy future.

PRODUCTION PLATFORMS

Crude oil production platforms are iconic structures that dot the world's oceans, marking the offshore frontier of the oil and gas industry. These mechanical wonders are deliberately positioned to tap into massive underwater resources, allowing crude oil production from underneath the seabed. Production platforms, which are designed for durability in severe marine conditions, play a critical role in supplying global energy demands. This investigation goes into the major characteristics of crude oil production platforms, offering light on their design, types, and relevance in offshore extraction.

TYPES OF PRODUCTION PLATFORMS:

1. Fixed Platform:

Fixed platforms are immovable structures that are securely anchored to the seafloor. They are often utilized in shallow waters with a rather stable substrate. Jacketed platforms, which feature steel legs that extend to the bottom and gravity-based constructions that sit on the seabed, are examples of fixed platforms.

2. Floating Production Platform:

Floating production platforms are adaptable constructions intended for use in deeper waters. Mooring systems or dynamic positioning can be used to secure them to the seafloor. Floating production, storage, and offloading (FPSO) boats, floating production systems (FPS), and tension leg platforms (TLPs) are examples of common types.

3. Compliant Tower:

Compliant towers are adaptable constructions that can endure lateral wind and wave stresses. They are ideal for moderate water depths and are frequently utilized in hurricane and typhoon-prone areas.

KEY COMPONENTS OF PRODUCTION PLATFORMS:

1. Topside Facilities:

Topside facilities are the upper portion of the platform that holds processing equipment, control rooms, and personnel housing quarters. They include oil and gas separation, water treatment, and power generation modules.

2. Substructure:

The substructure is the platform's lowest portion that offers stability and support. Jacket legs or a gravity-based foundation may be used for fixed platforms. A buoyant hull lies beneath the waterline of floating platforms.

3. Wellheads and Christmas Trees:

Wellheads are structures that are placed over individual wells, while Christmas trees are assemblages of valves and fittings that control the flow of oil and gas from the wells. These components are essential for well control and maintenance.

4. Risers and Pipelines:

Riser systems connect seafloor wells to topside facilities, allowing oil, gas, and other fluids to be transported. Pipelines connect offshore production to onshore processing and distribution facilities. They go from the platform to the shore.

5. Mooring Systems:

Mooring systems or dynamic positioning are used to keep floating platforms in place. Anchors and chains are often used to secure the platform to the bottom, maintaining stability under changing weather conditions.

SEPARATION FACILITIES

Crude oil collected from reservoirs is frequently mixed with other components such as oil, gas, and water. Efficient separation of these phases is a vital stage in crude oil processing. Separators, both two-phase and three-phase, are essential components in the oil and gas sector, designed to separate the mixture into distinct phases—oil, gas, and water. This in-depth examination delves into the principles, design concerns, and uses of two-phase and three-phase separators in the complicated world of crude oil processing.

TWO-PHASE SEPARATORS:

Principles of Two-Phase Separation:

Two-phase separators are generally used to separate oil and gas. The separation procedure is based on the density and volatility differences between the two phases.

1. **Gravity Separation:**

    The mixture of oil and gas enters the vessel of a two-phase separator, and due to gravity, the lighter gas phase rises to the top, while the heavier oil phase rests to the bottom.

2. **Gas Outlet and Liquid Outlet:**

    The separated gas is typically routed to the top of the separator and is collected through a gas outlet. The liquid oil is collected at the bottom and exits through a separate liquid outlet.

3. **Internal Components:**

    Internally, two-phase separators may have extra components such as baffles or coalescing devices to improve separation efficiency.

DESIGN CONSIDERATIONS FOR TWO-PHASE SEPARATORS

1. **Sizing:**

    Appropriate two-phase separator size is critical for effective separation. It takes into account factors such as oil and gas flow rates, required residence period, and vessel size.

2. **Inlet Design:**

   The intake design is critical for minimizing turbulence and ensuring the mixture enters the separator smoothly. This helps to prevent distinct stages from re-entrainment.

3. **Liquid Level Control:**

   It is critical to maintain an optimal liquid level for efficient separation. To regulate the liquid phase, liquid level control methods such as level controllers and dump valves are used.

4. **Gas Outlet Design:**

   The gas outlet must be designed to prevent liquid carryover into the gas stream. For this purpose, mist extractors or demisting devices are widely utilized.

THREE-PHASE SEPARATORS

Principles of Three-Phase Separation:

Three-phase separators are designed to handle a more complex mixture, involving oil, gas, and water. The separation process involves additional considerations to effectively separate the water phase from the oil and gas.

1. **Primary Separation:**

   Three-phase separators, like two-phase separators, achieve primary separation by allowing gas to rise to the top, oil to settle in the center, and water to collect at the bottom.

2. **Secondary Separation:**

   Secondary separation techniques are used to further separate the oil from the water. To improve separation efficiency, coalescing plates, corrugated plates, or other internals may be used.

3. **Oil Outlet, Water Outlet, and Gas Outlet:**

   Three-phase separators have separate oil, water, and gas outlets. The outlets are precisely placed to ensure that each phase is collected efficiently.

DESIGN CONSIDERATIONS FOR THREE-PHASE SEPARATORS

1. **Water Handling:**

   Managing the water phase is a fundamental part of three-phase separation. Considerations for design include water handling capacity, emulsion-breaking devices, and systems to prevent water carryover into the oil phase.

2. **Internal Components:**

   To maximize the separation process, three-phase separators may use advanced internals such as coalescers, cyclonic devices, or phase separation enhancers.

3. **Instrumentation:**

   For monitoring and managing the three-phase separation process, proper instrumentation, such as level sensors, pressure gauges, and temperature sensors, is essential.

4. **Residence Time:**

   A critical parameter is the residence time of the mixture within the separator. Adequate residence duration ensures complete phase separation and prevents carryover.

APPLICATIONS AND CHALLENGES

APPLICATIONS:

- **Upstream Processing:** At the wellhead, separators are used to handle the first mixture of oil, gas, and water.
- **Transportation:** Separated crude oil is less difficult to transport, lowering the danger of pipeline corrosion and hydrate formation.
- **Refining:** The separated components can then be routed to refining processes that are adapted to the individual properties of each phase.

CHALLENGES:

- **Emulsions:** Stable emulsions can hinder successful separation and may necessitate the application of Demulsifier.

- **Varying Flow Rates:** Flow rate and composition fluctuations make it difficult to maintain optimal separation efficiency.

- **Hydrate Formation:** Hydrates can build in cold conditions, reducing separator efficacy and necessitating further steps.

INNOVATIONS AND FUTURE TRENDS

1. **Advanced Internals:**

   Ongoing research focuses on producing more efficient internals, such as new coalescers and separators with increased phase separation capabilities.

2. **Digitalization and Automation:**

   By combining digital technologies and automation, separation processes may be monitored, controlled, and optimized in real time, increasing efficiency and lowering operating hazards.

3. **Hybrid Separation Technologies:**

   To address unique issues and optimize separation under varied situations, hybrid techniques integrating multiple separation technologies are being investigated.

4. **Separators:** Two-phase and three-phase separators are critical components in crude oil processing, separating the complicated mixture into discrete phases—oil, gas, and water. The efficiency of these separators has a direct impact on the extracted crude oil's quality as well as the overall safety and integrity of downstream processes.

5. **Gas Compression Stations**

- **Description:** In order to make natural gas transit through pipelines easier, gas compression facilities compress the gas.

- **Function:** Pumps natural gas to higher pressure for more economical transportation.

6. Oil and Gas Processing Plants:
- **Description:** Hydrocarbons are refined and treated in processing facilities to satisfy quality and market requirements.

- **Components:**
    - Distillation Units: Separate hydrocarbons based on boiling points.
    - Refining Towers: Refine crude oil into various products.

7. Storage Tanks:
- **Description:** Storage tanks are used to hold liquid natural gas, refined petroleum, or crude oil prior to processing or shipping.

- **Types:**
    - Floating Roof Tanks: Feature a floating roof that moves with the liquid level.
    - Fixed Roof Tanks: Have a permanent roof and are used for stable liquids.

8. Pumping Stations:
- **Description:** Pumping stations transport oil and gas through pipelines.

- **Types:**
    - Booster Stations: Increase pressure to overcome pipeline friction.
    - Mainline Stations: Move products over long distances through pipelines.

9. LNG Terminals:
- **Description:** Liquefied Natural Gas (LNG) terminals process and store natural gas in a liquid state for transportation.

- **Components:**
    - LNG Trains: Cool and liquefy natural gas.
    - Storage Tanks: Store LNG at very low temperatures.

10. **Water Injection Facilities:**
    - **Description:** Water injection facilities inject water into reservoirs to maintain pressure and increase oil recovery.
    - **Function:** Supports secondary or enhanced oil recovery methods.

11. **Flare Systems:**
    - **Description:** In order to stop dangerous gasses from escaping during crises or annual maintenance, flare systems burn off excess hydrocarbons.
    -
    - **Types:**
        - Elevated Flares: Positioned at a height to disperse combustion products.
        - Ground Flares: Located at ground level and burn off gases.

12. **Power Generation Units:**
    - **Description:** Power production units supply electricity for a range of tasks related to oilfield activities.
    - **Types:**
        - Gas Turbines: Use natural gas to generate power.
        - Diesel Generators: Use diesel fuel for power generation.

13. **Living Quarters (Living Quarters Platforms):**
    - **Description:** Personnel working on offshore installations are accommodated on living quarter's platforms.
    - **Features:**
        - Accommodation Modules: Living spaces for workers.
        - Recreation Areas: Facilities for relaxation and communal activities.

14. **Instrumentation and Control Systems:**
    - **Description:** Within oilfield facilities, a variety of processes are monitored and controlled via instrumentation and control systems.

- **Components:**
    - Sensors: Collect data on temperature, pressure, and flow.
    - Control Panels: Manage and adjust processes based on sensor data.

15. Water Treatment Facilities:
- **Description:** Facilities that treat water ensure that it satisfies environmental regulations before it is disposed of or used again.

- **Processes:**
    - Filtration: Removes impurities from water produced.
    - Chemical Treatment: Treats water with chemicals for purification.

Chapter Seven

OVERVIEW OF REFINING PROCESS

Crude oil refining is a complex and interrelated sequence of procedures that convert basic crude oil into a wide range of valuable products that power economies and industries all over the world. To meet rising energy demands, comply with environmental requirements, and embrace technological advancements that improve efficiency and sustainability, the sector is always evolving. With its many stages and sophisticated technologies, the refining process demonstrates the critical relationship between raw resources and the refined fuels and chemicals that power our contemporary world.

This in-depth examination looks into the complexities of crude oil refining, exploring each stage of the process, the technologies involved, and the significance of the refined products.

1. DISTILLATION: THE INITIAL SEPARATION PROCESS

Distillation is the first phase in the refining of crude oil, focused on the separation of hydrocarbons depending on their boiling points. Crude oil is a complicated mixture of hydrocarbons with variable molecular weights and characteristics that comes from the earth. Distillation uses boiling point differences to separate these components into discrete fractions.

Process:

1. **Fractional Distillation:** The crude oil is heated in a furnace, and the resultant fumes are sent into a distillation column. This column is made up of many trays or packaging materials. As the vapors climb through the column, they cool, and different hydrocarbons condense into liquid fractions at different heights.

2. **Fraction Separation:** The hydrocarbons are separated into fractions by the column depending on their boiling points. Lighter fractions, such as gasses and naphtha, rise to the surface and are collected at higher levels, whilst heavier fractions, such as diesel and residual oil, settle at lower levels.

3. **Condensation and Collection:** Each fraction is condensed back into a liquid condition, and the separated products are collected in various trays or levels of the distillation column.

Key Considerations:

- **Crude Oil Characteristics:** The presence of light and heavy hydrocarbons in crude oil affects the design and efficiency of the distillation process.

- **Column Efficiency:** The design and operation of the distillation column impact the separation efficiency, with the column height and internals playing crucial roles.

2. CONVERSION PROCESSES: (ENHANCING PRODUCT YIELDS)

Overview:

Conversion techniques are used to change the molecular structure of specific hydrocarbons in order to increase the production of valuable products such as gasoline and diesel. These techniques are critical for increasing the output of high-demand fuels.

Process:

1. **Cracking:** Cracking is the process of converting big hydrocarbons into smaller, more valuable ones. Common methods include catalytic cracking and thermal cracking. Catalytic cracking uses a catalyst to help the process along, whereas thermal cracking relies on high temperatures.

2. **Hydrocracking:** Hydrocracking is the combination of cracking and hydrogen under high pressure and temperature. This method is useful for converting heavy hydrocarbons into lighter, more value compounds.

Key Considerations:

- **Product Flexibility:** Conversion procedures provide flexibility in the manufacture of a wide range of products, allowing refineries to modify production in response to market demands.
- **Catalyst Selection:** Catalyst selection is critical in catalytic cracking because it influences reaction rates and product selectivity.
- **Hydrogen Availability:** Hydrocracking is dependent on hydrogen supply; therefore refineries frequently have hydrogen generation facilities to support this process.

-

3. TREATMENT PROCESSES: IMPROVING PRODUCT QUALITY

Overview:

Treatment techniques attempt to increase refined product quality by removing impurities and undesired components. Hydro treating is a critical treatment step in crude oil refining.

Process:

1. **Hydro treating:** The use of hydrogen to remove contaminants such as sulfur, nitrogen, and metals from refined products is known as hydro treating. Catalysts, which are often made of metals such as nickel and molybdenum, aid in the conversion of contaminants into more stable molecules.
2. **Desulfurization:** Desulfurization is used to remove sulfur from fuels in order to meet environmental laws and provide cleaner-burning fuels.

Key Considerations:

- **Environmental Compliance:** Treatment methods are essential for achieving environmental criteria by lowering sulfur content in fuels.
- **Catalyst Regeneration:** Hydro treating catalysts may require regeneration to maintain their effectiveness, and refineries frequently have dedicated regeneration units.

4. SEPARATION PROCESSES: REFINING SPECIFIC PRODUCT STREAMS

Overview:

In order to guarantee that certain product streams fulfill exact specifications and raise the general caliber of refined products, separation processes further refine particular product streams.

Process:

1. **Dehydration:** The process of taking the water out of refined products is called dehydration. This is essential to avoiding corrosion and problems with stability in the finished goods.

2. **Fractionation:** A separation technique called fractionation is used to refine particular product streams. In order to produce products with exact specifications, mixed hydrocarbons must be separated into different fractions.

Key Considerations:

- **Water Content:** In the storage and transportation of refined products, dehydration is crucial to avoiding water-related problems.
- **Fractionation Efficiency:** To guarantee accurate hydrocarbon separation into the desired fractions, fractionation facilities need to be well-designed and operated efficiently.

5. FINAL BLENDING: ACHIEVING DESIRED PRODUCT SPECIFICATIONS

Overview:

In order to attain the desired qualities and match market criteria, different refined products are combined in particular ratios during the final blending stage. Producing goods that satisfy both consumer demands and legal regulations requires blending.

Process:

1. **Blending Tanks:** Blending tanks are used to combine several streams of refined goods, such as jet fuel, diesel, and gasoline.
2. **Product Specification Control:** The ratios of various ingredients are meticulously regulated throughout blending processes to meet

the required standards for particular fuel grades and other refined products.

Key Considerations:
- **Quality Control:** Strict quality control procedures are needed in blending operations to guarantee that the finished goods live up to consumer expectations and regulatory requirements.
- **Flexibility:** Refineries are frequently able to modify their blending ratios in response to seasonal fluctuations and market demands.

6. ENVIRONMENTAL CONTROLS: MITIGATING ENVIRONMENTAL IMPACT

Overview:

Throughout the refining process, environmental controls are included to limit the influence on water and air quality. These controls are necessary to guarantee adherence to environmental laws and promote environmentally friendly activities.

Process:
1. **Flare Systems:** By burning surplus gases, flare systems help cut down on the amount of volatile organic compounds (VOCs) and other pollutants released into the atmosphere.
2. **Wastewater Treatment:** Before being released into the environment, wastewater from different refining processes is treated to remove impurities.

Key Considerations:
- **Regulatory Compliance:** Environmental control systems are designed and operated with compliance in mind, as refineries are subject to strict environmental rules.
- **Sustainable Practices:** Within the sector, there is a rising emphasis on implementing sustainable practices, like cutting back on flaring and using cutting edge technologies for wastewater treatment.

7. PRODUCT DISTRIBUTION: DELIVERING TO END USERS

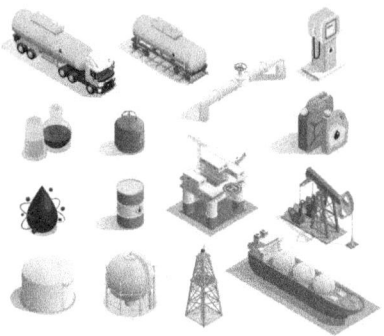

Overview:

Tankers, pipelines, and other delivery methods are used to deliver the finished refined goods to end customers. This phase is essential for guaranteeing customers and other sectors a consistent supply of refined products.

Process:

1. **Storage Tanks:** Before being sent to distribution centers, refined products are kept in storage tanks.

2. **Transportation:** Refined products are transported by a variety of means, such as trucks, ships, and pipelines, to final consumers. Pipeline systems are an economical and effective way to move bulk materials over great distances, but ships and trucks provide you more options when it comes to reaching other markets.

Refineries place storage tanks in a strategic location close to distribution centers to enable prompt delivery and uphold a stable supply chain. By acting as reservoirs for a range of refined products, these tanks enable refineries to adjust to changes in demand and seasonal swings. The distribution infrastructure makes sure that a dependable and varied selection of refined products is available to end consumers, which include homes, companies, and transportation networks.

In brief, the refining operations can be classified into following major heads:

1) Fractionation also known as distillation, this process separates crude oil in atmospheric and vacuum distillation towers into hydrocarbon compounds of varying boiling points ranges called "fractions" or "cuts".

2) Conversion Processes are used to change structure of molecules by Cracking (thermal and catalytic), Alkylation, Polymerization, and Isomerization & Catalytic reforming.

3) Treatment Processes are used for additional processing to prepare final finished products and may include:-

    - Separation/removal of aromatics & Naphthenes
    - Removal of impurities
    - Chemical or physical separation
    - Desalting
    - Hydrodesulphurization
    - Sweetening
    - Solvent extraction
    - Solvent DE waxing
    - Solvent refining

4) Blending by mixing and combining hydrocarbon fractions, additives, and other components to produce finished product with desirable performance properties.

5) Other operations including but not limited to
    - Sour-water stripping
    - Waste water disposal, oil water separator
    - Cooling, storage & handling systems
    - Hydrogen plants
    - Sulfur recovery units
    - Acid treatment
    - Generators & firefighting systems

Chapter Eight

TYPES OF REFINERIES

There are 3 types of refineries based on the distillate fractions processed by them:
1. Skimming plants
2. Lubricating and wax plants
3. Complete run-down refineries

Skimming Plants

Figure 7: Skimming Plant (Refinery)

Skimming plants, sometimes referred to as topping plants, are a particular kind of petroleum refining plant used for crude oil's early processing. These facilities specialize in atmospheric distillation, also referred to as topping, which is the major method of separating crude oil into its constituent parts. "Skimming" describes the process of removing the lighter fractions from crude oil by "skimming off" them according to their various boiling temperatures.

Crude oil is heated in a furnace at a skimming plant, and the heated vapors are then sent into a distillation column. Fractional distillation is used in this column to separate the crude oil into liquid fractions at varying heights depending on the boiling points of the individual hydrocarbons. Diesel and residual oil settle at lower depths, while lighter fractions like gasses, naphtha, and kerosene rise to the top and are collected at higher levels.

Skimming plants are distinguished by their ease of use and effectiveness in dissolving crude oil into its constituent parts. Even though their main output is lighter stuff like naphtha and gasoline, the heavier fractions are sometimes transported to secondary refining facilities for additional processing. Skimming facilities are the first stage of the intricate refining process, helping to produce petrochemicals and vital fuels that power a variety of industries and satisfy the wide range of energy demands of society.

Although precise details about small-scale skimming plants may differ, the following are instances of compact refining establishments that generally integrate atmospheric distillation or skimming units:

1. **Cheyenne Refinery (Wyoming Refining Company) - USA:**
   - The smaller Cheyenne Refinery specializes on the processing of crude oil. An atmospheric distillation unit is part of it, which divides crude oil into its component parts.

2. **Pertamina Balongan Refinery - Indonesia:**
   - Serving the local market, Pertamina runs the smaller-scale Balongan Refinery, which has the capacity to perform atmospheric distillation.

3. **Nayara Energy Refinery - India:**
   - Although not very modest, the Vadinar Refinery in India owned by Nayara Energy contains refining facilities that use air distillation for primary separation.

4. **Petromidia Refinery (Rompetrol) - Romania:**
    - The Petromidia Refinery, which is part of the Rompetrol Group, is a Romanian refinery with a smaller capacity. It includes units for atmospheric distillation.

Please keep in mind that the definition of "small" varies, and these refineries may nonetheless have large capabilities for the regions they service. Furthermore, refinery facilities are frequently expanded or modified, which might have an impact on their capacity over time. Official statements from refining businesses or industry studies are preferred for the most accurate and up-to-date information.

Lubricating & Wax Plants

Lubricating and wax plants are a subset of the refining sector that manufactures critical items such as lubricants and waxes. These facilities are critical in addressing the varying needs of numerous industries, ranging from automotive and aviation to manufacturing and consumer goods. This investigation digs into the diverse processes and products linked with lubricating and wax plants, highlighting their importance in defining current applications.

These plants start with Crude oil and get the same products as Skimming plants, and further break down the Fuel oil into equal parts of Paraffin distillate and Cylinder stock. Lubricating oil and Paraffin wax are obtained from Paraffin distillate, whereas the Cylinder stock is sold commercially.

Lubricating Plants:

Overview:

Lubricating oil is a necessary component for reducing friction and wear between moving elements in machinery and motors. Lubricating facilities specialize in refining crude oil to provide a variety of lubricants with varied viscosity, additives, and qualities that are customized to specific purposes.

Key Processes:

1. **Hydro treating:** This process removes contaminants and sulphur from the base oil, hence improving its stability and performance.
2. **Additive Blending:** This process introduces particular additives to increase lubricant qualities such as anti-wear, anti-oxidation, and viscosity index.

Examples:

1. **Mobil 1 Lubricating Oil Plant (ExxonMobil) - USA:**
   - ExxonMobil produces high-performance lubricants, such as the well-known Mobil 1 synthetic motor oil, in a state-of-the-art lubricating oil plant.
2. **Castrol Lubricants Plant (BP) - United Kingdom:**
   - BP's Castrol lubricants plant manufactures a variety of lubricants for automotive, industrial, and marine applications.
3. **Shell Lubricants Blending Plant - Singapore:**
   - Shell has lubricant blending plants around the world, including one in Singapore, where a variety of lubricant products are developed.

Wax Plants:

Waxes are versatile materials utilized in a wide range of industries for everything from candles to industrial procedures. Wax plants refine crude oil or natural gas to extract waxes, each with unique properties based on the intended application.

Key Processes:

1. **Solvent DE waxing:** Solvents are used to extract heavy waxy components from oil, resulting in desired paraffin wax.
2. **Hydrocracking:** Using heat and pressure, this converts heavy fractions into lighter, more valuable waxes.

Examples:

1. **Sasol Wax Manufacturing Plant - South Africa:**

    Sasol, a worldwide energy and chemical firm, owns and runs wax manufacturing factories that produce a variety of synthetic and petroleum-based waxes.

2. **Petro-Canada Lubricants and Specialty Products - Canada:**
    - Suncor Energy subsidiary of Petro-Canada has a specialty products unit that manufactures a variety of waxes for industrial purposes.

3. **Holly Frontier Wax Plant - USA:**
    - Holly Frontier operates wax factories in the United States through its subsidiaries, contributing to the manufacturing of a variety of wax products.

Significance of Lubricating and Wax Plants:

1. **Industrial Applications:**

    o Lubricants are necessary in machinery and engines to ensure smooth functioning and to increase the life of the equipment.

    o Waxes are used in a variety of sectors, including packaging, cosmetics, and pharmaceuticals.

2. **Advanced Formulations:**

    o Lubricating plants invest on R&D to produce sophisticated synthetic lubricants with better performance attributes.

3. **Specialized Markets:**

    o Lubricating and wax plants serve niche markets, creating goods customized to meet the particular needs of several sectors.

4. **Global Presence:**
    - Large energy corporations run wax and lubricating factories all around the world, which helps to maintain a steady and varied supply chain for these necessary goods.

5. **Innovation and Sustainability:**
    - Constant innovation in lubricant compositions aims to meet changing industry standards, minimize environmental impact, and increase efficiency.
    - Environmental responsibility is enhanced via sustainable wax production methods, which include lowering energy use and emissions.

To sum up, lubricating and wax factories are essential to the production of essential items that improve industrial processes, lubricate machinery, and serve as building blocks for a wide range of applications.

Figure 8: Complete Run down Refinery

Complete run-down Refinery

A sophisticated structure built to convert crude oil derived from the Earth into usable goods that power our daily lives is a complete derelict refinery. A comprehensive rundown facility, in contrast to more basic refineries, is outfitted with multiple units that function as a well-composed symphony to extract valuable components from crude oil.

A complete rundown refinery is a complicated facility meant to convert crude oil derived from the Earth into usable goods that fuel our everyday life. A comprehensive rundown facility, as opposed to a simpler refinery, is outfitted with multiple units that operate together like a well-orchestrated symphony to extract valuable components from crude oil.

Crude oil passes through a sequence of harmonious transformations in this huge refinery symphony, with each unit playing an important role in producing the fuels and materials that power our world. The refinery choreographs a dance that transforms ancient organic relics into the energy sources and goods we rely on every day, from the initial separation in air distillation to the molecular rearrangements in cracking units.

This refinery also starts with Crude oil and makes all the products that the skimming plant and the lubricating & wax plant make. In addition to these products, it makes Bright stock and Petrolatum, from which many more products are produced.

Gasoline made by the distillation of crude oil in straight run (complete) refineries is called straight run gasoline and for many years was the only kind of gasoline manufactured. It is a fact; however, that the greater part of the gasoline marketed today is blended. Most of the blended gasoline is preferable to the straight run products particularly if the added part is ethanol (Ethyl alcohol).

In the first distillation of crude oil, crude naphtha is the first fraction to vaporize, and when this is re-distilled it breaks up into naphtha, benzene and gasoline.

Naphtha however lacks the low burning constituents essential to a good motor fuel while gasoline has too much—therefore a blend of naphtha and gasoline makes a high-quality motor fuel. To reduce knocking of the engines and improve octane rating, small amount of an additive "Tetra ethyl lead" is added to the gasoline.

Typical Crude oil distillation and downstream refinery processing

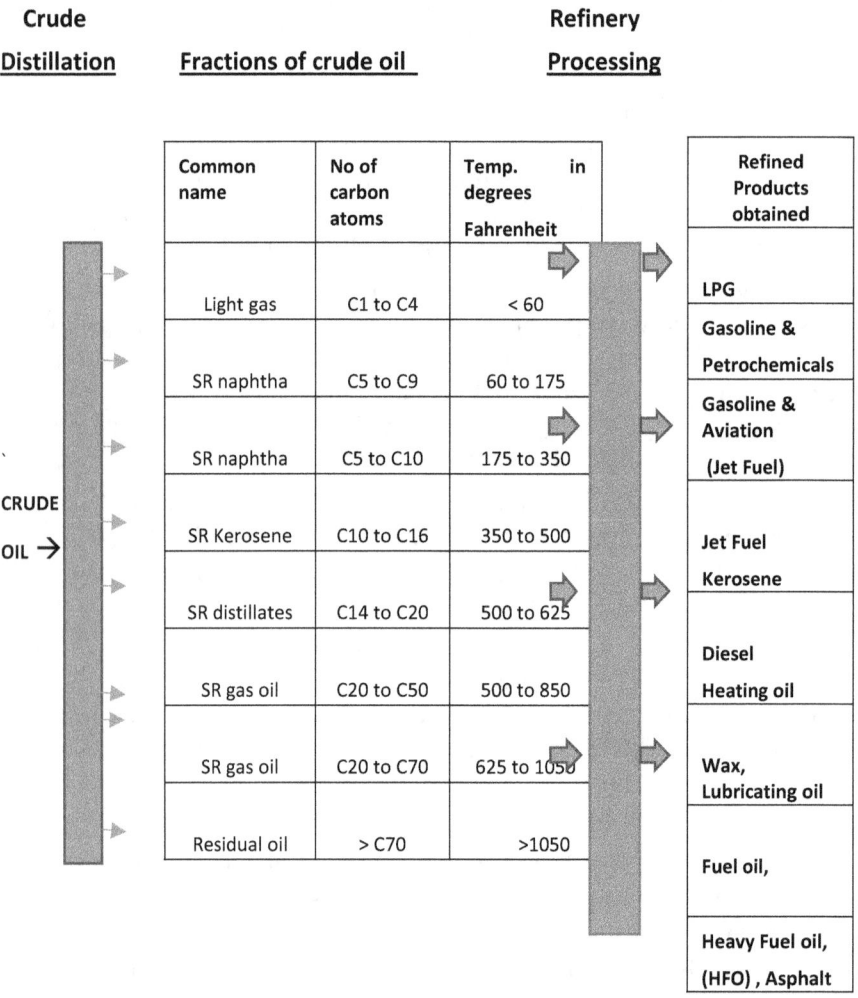

Figure 9: *Typical Crude Distillation & Downstream products*

In another classification used in some countries the refineries are classified as "Topping", "Hydro-skimming", "Conversion" or "Deep conversion" refineries. The refinery's processing varies based on type of crude, location and equipment used.

For ease of comparison and understanding, the processes carried out by these 4 types of refineries are enumerated in table below (figures may have minor variation based on refinery process):

S. No.	Process / output	Topping refinery	Hydro-skimming refinery	Conversion refinery	Deep conversion refinery
1	Crude distillation	Yes	Yes	Yes	Yes
2	Capability to alter yield	No	Yes	Yes	Yes
3	Control sulfur levels	No	Yes	Yes	Yes
4	Catalytic reforming	No	Yes	Yes	Yes
5	Hydro treating	No	Yes	Yes	Yes
6	Blending	No	Yes	Yes	Yes
7	Catalytic cracking	No	No	Yes	Yes
8	Hydrocracking	No	No	Yes	Yes
9	Alkylation	No	No	Yes	Yes
10	Coking	No	No	No	Yes
11	Gasoline yield (vol. %)	>30	25 to 30	40 to 45	46 to 48
12	Diesel & Aviation fuel yield (vol. %)	30	30	33	>40

Figure 10: Product comparison of 4 types of refineries

Topping Refinery

A topping refinery's main goal is to separate crude oil into its constituent fractions, each of which has a different boiling point. The method used to achieve this separation is called atmospheric distillation, and it involves heating crude oil in a tall column called a distillation tower or atmospheric column. Different hydrocarbon fractions condense at different levels as a result of the steady temperature decrease that occurs as the crude oil ascends down the column.

The distillation tower is made up of several trays or packing materials and is frequently a massive building that dominates the refinery skyline. With the lightest gases, like methane and propane, at the top and heavier products, like diesel and residual fuel oil, at the bottom, these elements help separate crude oil into fractions. A range of products with different boiling points are the end result, and these can be processed further in units farther down the line.re decrease, causing certain hydrocarbon components to condense at particular concentrations.

The efficiency and simplicity of topping refineries define them. Their strategic importance cannot be emphasized, even if they are not as complex as more sophisticated refining units. For areas where there is a need for basic refined products but no requirement for large secondary processing units, topping refineries are a cost-effective option due to the simple nature of the distillation process.

Naphtha is a highly adaptable feedstock for petrochemical companies and is one of the main products obtained by topping refineries. Naphtha is used to make a number of compounds, including as synthetic fibers and plastics. The light gases that are recovered during the distillation process also contribute to the energy infrastructure by being essential parts for both industrial and residential use.

The output products include:

- Light gas
- Naphtha

- Kerosene, aviation fuel
- Diesel
- Heating oil
- Heavy fuel oil
- In some cases, low octane gasoline

Hydroskimming Refinery

These refineries have all the functions of Topping refineries and in addition have Catalytic reforming, hydro treating and blending facilities. They have processes to control Sulfur content of refined products and also upgrade naphtha to gasoline by catalytic reforming to meet octane specifications. Hence, they are able to process high sulfur crude oils.

The hydro skimmer is the central component of the hydro skimming refinery; it combines air distillation with a simple hydro treating unit. The principal aim of these refineries is to separate crude oil into valuable fractions and at the same time reduce the sulfur content by hydro treating. Refineries with more complexity might use more processing units to extract higher-value products, but hydro skimming refineries balance things out by concentrating on what matters most.

A hydro skimming refinery's atmospheric distillation procedure is comparable to that of a conventional topping unit. After heating, crude oil goes through fractionation as it moves up the distillation column, separating into heavier residues, gases, kerosene, and diesel. But the addition of a hydro treating unit after the distillation column is what distinguishes hydro skimming refineries.

By reducing sulfur, nitrogen, and other contaminants in the fractions using a catalytic process called hydro treating, cleaner and more ecologically friendly products are produced. Hydro treating in hydro skimming refineries usually focuses on reducing sulfur in diesel and other heavy fractions. By taking this extra step, the refined goods are of higher quality and meet consumer demands for cleaner fuels as well as stricter environmental requirements.

These refineries are frequently preferred in areas where there is a need for an affordable solution and a moderate demand for more advanced refined goods. When compared to more complicated refineries, hydro skimming units are a more cost-effective solution due to their simplicity, which enables faster construction and operational start-up.

Refineries that use hydro skimming technology offer a practical method of refining crude oil, but they are not without difficulties. The spectrum of refined products is limited as a trade-off for simplicity. These refineries mostly produce basic fuels, and in comparison to more adaptable refining setups, their capacity to adjust to shifting market demands may be limited.

Conversion Refinery

Within the ever-changing world of petroleum refining, conversion refineries are essential to the complex process of turning crude oil into a range of valuable products. Conversion refineries are distinguished from simpler topping or hydro skimming refineries by having sophisticated processing systems that go beyond simple distillation. These refineries use a variety of conversion methods in order to improve product yields, upgrade lower-value fractions, and provide a wide variety of valuable end products.

These refineries have all the functions of Hydro skimming refineries and in addition have Catalytic cracking or hydrocracking by which they can convert heavy crude oil fractions to gas oils with high yield. They are able to get the light streams of gasoline, diesel, aviation fuel and feedstock for petrochemicals as per market demand. In spite of good yield, they still produce some low value products like asphalt and residual fuel.

Important Elements in a Conversion Refinery:

1. Units for Catalytic Cracking:
Catalytic cracking units, which are the heart of conversion refineries, use catalysts to convert heavy hydrocarbons into lighter, more valuable products. Fuel and light olefin production are improved via the commonly utilized procedure known as fluid catalytic cracking (FCC). Conversion

refineries increase the production of lighter hydrocarbons that are in great demand by breaking up bigger molecules.

2. **Hydrocracking Units:** Using hydrogen to break down complex hydrocarbons, hydrocracking is an essential conversion process. This method works especially well for turning heavy fractions into useful products, such as jet fuel and diesel. By increasing the yield of middle-distillate products with better qualities, hydrocracking satisfies strict quality requirements and meets the growing demand for cleaner fuels.

3. **Coking Units:** Petroleum coke, gasoline, and distillate fuels are among the important products that are produced from heavy residues by coking units, which include delayed cokers and fluid cokers. Processing heavy crude oils and wastes that would be difficult to handle in less complex refining settings is made possible by coking.

4. **Alkylation Units:** Alkylation units mix light olefins and isobutane to produce high-octane gasoline. By using less environmentally hazardous components, this procedure improves gasoline's qualities and qualifies it for use in high-performance engines while also lessening its impact on the environment.

5. **Isomerization Units:** The process of isomerization involves the transformation of straight-chain hydrocarbons into their branched counterparts. By improving gasoline's combustion properties and octane rating, this method helps to produce fuels that burn cleaner and more efficiently.

The Function of Conversion Refineries

1. **Maximizing Value from Crude Oil:** The purpose of conversion refineries is to turn heavier, lower-value fractions into products with higher market demand and value in order to maximize the value extracted from crude oil.

2. **Adaptability in the Choice of Crude:** These refineries can handle a wide range of crude oils, including sour and heavy grades, with flexibility, enabling strategic changes based on crude availability and market conditions.

3. **Fulfilling Strict Environmental Requirements:** These refineries' conversion operations help them meet strict environmental criteria by

generating low-sulfur fuels and other cleaner, greener products.

4. Taking Care of Market Demands: Refineries for conversion can adjust to changing market conditions. These refineries can adapt their operations to create the necessary product mix when consumer tastes and regulatory requirements change.

5. Enhancing Energy Security: Conversion refineries contribute to improved energy security by diversifying their product portfolio and providing a variety of fuels and petrochemical feedstock.

Deep Conversion Refinery

Within the petroleum refining industry, a deep conversion refinery is a highly developed and complex plant. It stands out as a technological marvel made to turn the heavier, less attractive components of crude oil into profitable end products in order to maximize value extraction. Deep conversion refineries are designed to handle the most difficult and heavy components of crude oil, pushing the limits of traditional refineries' capabilities.

After being extracted, crude oil comprises a variety of hydrocarbons, from light gasses to heavy leftovers. Deep conversion refineries are dedicated to handling the heavier components that are frequently disregarded, whereas standard refineries concentrate on sorting and processing the lighter fractions. To turn these heavier fractions into high-value products, such as vacuum gas oil, heavy gas oil, and residues, they go through a number of complex procedures.

These are special class of refineries also known as coking refineries. They have hydrocracking and catalytic cracking to get gas oil, and also coking units which convert the least valuable residual oil into lighter streams to produce more valuable light products.

Most of the refineries in Europe and Japan are either Hydro skimming or Conversion refineries. On the contrary almost all refineries in US and Canada are Conversion or Deep conversion refineries.

KEY PROCESSES IN DEEP CONVERSION:

1. Hydrocracking:

The hydrocracking unit, which converts big hydrocarbons into smaller, more valuable ones, is the central component of a deep conversion refinery. This is accomplished by adding hydrogen, which is catalyzed by unique catalysts, at high pressure and temperature. In addition to yielding useful products like jet fuel and diesel, hydrocracking lowers sulfur content to comply with strict environmental regulations.

2. Visbreaking:

Another crucial step in deep conversion is visbreaking. Its process entails using heat to convert heavy wastes into lighter products. This helps to reduce the viscosity of the leftovers, which makes them easier to handle in later refining procedures, in addition to producing more valuable components.

3. Coking:

Coking units, such as fluid or delayed coking, are frequently used in deep conversion refineries. Petroleum coke and valuable lighter products are produced by thermally breaking down the heaviest components. The coke finds uses in the steel and aluminum industries, while having less value as fuel.

4. Residue Desulfurization:

One essential component of deep conversion is sulfur removal. Residues are desulfurized using specialized machines, guaranteeing that the finished products adhere to strict environmental laws. In order to minimize the impact on the environment and produce cleaner fuels, this step is essential.

SIGNIFICANCE OF DEEP CONVERSION:

1. **Maximizing Value from Crude Oil:**

    By adding value to the heavier, frequently unused fractions of crude oil, deep conversion refineries enable a more thorough and effective use of this valuable resource.

2. **Producing High-Value Products:**

   Deep conversion produces high-value end products such as specialty chemicals, lubricants, and cleaner fuels through its sophisticated procedures.

3. **Environmental Compliance:**

   Deep conversion refineries support international initiatives to reduce air pollution and enhance air quality by generating cleaner fuels and lowering the sulfur level of their products.

4. **Enhanced Energy Security:**

   Deep conversion increases a nation's or region's energy security by maximizing the output of necessary goods from locally produced crude oil, lowering reliance on outside sources.

EXAMPLES OF DEEP CONVERSION REFINERIES:

1. **Petrobras Refinery (Rnest) - Brazil:**

   One deep conversion refinery in Brazil that uses cutting edge technology to treat heavy crude oil is Petrobras' Refinaria Abreu e Lima (Rnest).

2. **SK Energy Ulsan Refinery - South Korea:**
   - The SK Energy-run Ulsan Refinery is renowned for its deep conversion capabilities, effectively processing heavy crude oil to suit South Korea's energy needs.
   -

3. **Indian Oil Paradip Refinery - India:**

   - Deep conversion units at the Indian Oil Corporation-run Paradip Refinery increase India's capability for refining and enable the production of premium fuels and products.

Deep conversion refineries are trailblazers in the complex field of petroleum refining, stretching the limits of what can be done with crude oil. Their cutting-edge methods not only maximize value from each barrel but also help create a more sustainable and clean energy environment. Deep conversion refineries are essential to ensure that the enormous resources stored within crude oil are effectively utilized in order to fulfill the demands of a society that is both dynamic and energy-hungry as it moves toward greener energy alternatives.

Chapter Nine

BASIC CHEMISTRY

The characteristic properties of any substance depend on the arrangement of atoms composing its molecules.

A molecule is the smallest possible unit of a pure substance still possessing the characteristic properties of the substance. On further division a molecule disintegrates into atoms of its elements. For example, 1 molecule of water i.e., H_2O, H-O-H, contains 2 atoms of hydrogen and 1 atom of Oxygen.

Crude oils are mixtures of many hydrocarbons, mainly compounds of carbon (C) and hydrogen (H). Other elements are negligible and present in very small quantities only. Sulfur (S) has important effect in product quality. Various petroleum products like gas, gasoline, naphtha, gas oil, fuel oil, heavy fuel oil, lubricants, wax and bitumen are manufactured from Crude oil.

HYDROCARBONS

Hydrocarbons may be solid, liquid or gas at normal temperature and pressure depending upon the arrangement of carbon atoms in their molecules as under:

- Hydrocarbons with up to 4 carbon atoms: Gaseous
- Hydrocarbons with 5 to 19 carbon atoms: Liquid
- Hydrocarbons with 20 or more carbon atoms: Solid

Some examples of hydrocarbons: Methane, Pentane, Cyclopentane, Ethylene, Acetylene, Benzene, Toluene, Naphthalene etc.

NON-HYDROCARBONS

Crude oils contain minor quantities of non-hydrocarbons, most important being Sulfur (S), Nitrogen (N) and Oxygen (O). In some crude oils there are small amounts of Sodium (Na), Potassium (K), Vanadium (Vn) or Nickel (Ni).

Some examples of non-hydrocarbons: Ethyl mercaptan (C_2S_5SH), Diethyl sulphide ($(C_2H_5)_2S$), Thiophene (C_4H_4S), Phenol, Cresol, Xylenol, Naphthenic acids etc.

HYDROCARBON REACTIONS

In crude oil refining certain basic reactions play important role. The following are most important of these reactions:

CRACKING:

It is the process of breaking large hydrocarbon molecules into smaller molecules by the application of heat and breaking the Carbon–Carbon bonds.

$$CH_3\text{-}CH_2\text{-}CH_2\text{-}CH_2\text{-}CH_3 \rightarrow CH_2\text{-}CH\text{-}CH_3 + CH_2\text{=}CH\text{-}CH_2\text{-}CH_3$$

$$C_5H_{12} \rightarrow C_3H_6 + C_4H_8$$
$$\text{Pentane} \qquad \text{Propylene} \quad \text{Butylene}$$

DEHYDROGENATION:

It is the process of eliminating hydrogen atoms from a molecule.

$$C_2H_6 \rightarrow C_2H_4 + H_2$$
Ethane   Ethylene   Hydrogen

HYDROGENATION:

It is the reverse of dehydrogenation i.e., adding of hydrogen atoms to a molecule.

$$C_2H_4 + H_2 \rightarrow C_2H_6$$
Ethylene   Hydrogen   Ethane

PYROLYSIS:

It is a severe form of thermal cracking and is usually accompanied by rearrangement of fragments. For Example, methane heated to 1200 degrees Centigrade yields Hydrogen, Carbon and Acetylene

$$3\ CH_4 \rightarrow 5\ H_2 + C + CH{=}CH$$

Methane   Hydrogen Carbon Acetylene
(Heated to 1200 deg C)

CYCLIZATION:

It is the process of converting a chain molecule into a ring molecule. During this process hydrogen is released or lost from the chain molecule.

$$CH_3\text{-}CH_2\text{-}CH_2\text{-}CH_2\text{-}CH_2\text{-}CH_3 \rightarrow C_6H_{12} + H_2$$

n-Hexane   cyclo-hexane hydrogen

ALKYLATION:

In this process the light olefins (mainly C3 and C4) combine with iso-butane to produce high octane gasoline blend stock, also known as alkylate. The FCC units are the key suppliers of iso-butane and light olefins; hence alkylation units are found in those refineries which have FCC units.

Alkylation is the process of introducing a straight or branched chain hydrocarbon group or an alkyl group into an aromatic or branched chain hydrocarbon. Depending on the process either hydrofluoric acid (HF) or sulphuric acid (H_2SO_4) is used as catalyst. Utmost care needs to be taken in the handling, storage and transportation of hydrofluoric acid and sulphuric acid because of environmental and health hazard posed by these strong acids.

$$C_6H_6 \quad + \quad C_8H_{18} \rightarrow C_{14}H_{24}$$

Benzene   octane   octyl benzene

Also

$$\underset{\text{Isobutane}}{\begin{array}{c} CH_3 \quad CH_3 \\ | \quad\quad | \\ CH_3\text{-}CH \\ | \\ CH_3 \end{array}} + \underset{\text{Propylene}}{\begin{array}{c} CH_3 \\ | \\ CH \\ || \\ CH_2 \end{array}} \rightarrow \underset{\text{Isoheptane}}{\begin{array}{c} CH_3 \quad CH_3 \\ | \quad\quad | \\ CH_3\text{----}C\text{--------}CH \\ | \quad\quad | \\ CH_3 \quad CH_3 \end{array}}$$

Alkylation results in production of premium gasoline blend stock because the alkylate contains no sulfur and no aromatics.

ISOMERIZATION:

The process of isomerization rearranges the low octane C5 and C6 normal Paraffins in light straight run naphtha to produce higher octane C5 and C6 iso-Paraffins resulting in significant increase of octane of resulting naphtha stream (isomerate). Also, the product obtained is free from sulfur and benzene.

POLYMERIZATION OR COPOLYMERIZATION:

Polymerization is the process of combination of a number of unsaturated molecules of the same or different compounds to form single large molecule call the polymer. During the process two or three light olefin molecules (C3 or C4) combine to produce high octane olefinic gasoline blend stock. Polymers are generally plastics the properties of which depend on their molecular size.

$$n\text{-}CH_2\text{=}CH_2 \rightarrow CH_3\text{-}CH_2\text{-}CH_2\text{-}CH_2\text{-}CH_2\text{-------}CH_3$$

ethylene   polyethylene

OXIDATION:

It is the reaction of oxygen with a molecule that may or may not already contain oxygen. Oxidation may be partial resulting in the addition of oxygen into the molecule or in the elimination of hydrogen from it or it may be complete, forming carbon dioxide and water.

Partial oxidation

$$2CH_3CH_2OH + O_2 \rightarrow 2\,CH_3\text{-}CHO + 2H_2O$$

ethyl alcohol   oxygen   acetaldehyde   water

Complete oxidation

$$CH_4 + 2O_2 \rightarrow CO_2 + 2H_2O$$

methane   oxygen   carbon dioxide   water

ETHERIFICATION:

This process combines C4 and/or C5 olefins with methanol or ethanol to produce ether, which is premium gasoline blend stock with high octane.

The most common etherification combines methanol with iso-butene to produce methyl tertiary butyl ether (MTBE). Similarly, ETBE (ethyl tertiary butyl ether) is made from ethanol and iso-butene and TAME (tertiary amyl methyl ether) is made from methanol and iso-amylene.

HYDRODESULFURIZATION:

It is the elimination of Sulfur from Sulfur containing chain molecules in crudes or distillate by the action of hydrogen under pressure over a catalyst.

$$C_{18}H_{33}SH + H_2 \rightarrow C_{16}H_{34} + H_2S$$

ULSD (Ultra low sulfur diesel) production needs severe desulfurization of diesel blend stock which is achieved by hydro treating.

CATALYSIS:

It is the alteration of the rate of a chemical reaction by the presence of foreign substance called catalyst that remains unchanged at the end of the reaction. For instance, hydrogenation is carried out using metallic platinum or nickel and cracking of hydrocarbons using a silicate. A common example is preparation of vegetable ghee by hydrogenation.

Chapter – Ten

DISTILLATION AND FRACTIONATION

The refining process is very simple, mainly based on fractional distillation at various temperatures. Petroleum generally called "CRUDE OIL," is made up of various hydrocarbons, having different vaporizing temperatures, which is the basis of refining industry. More than 2500 types of refined products are obtained by refining & processing of the crude oil.

The refinery activities begin with receipt of crude, refining, storage and shipping of finished products to Terminals through Rail, Road, Ship or Pipeline network.

In general, the refinery processing mainly uses the following:

- Still
- Cooling tank
- Tail house
- Pipelines
- Pumps & Motors
- Storage tanks

The market requires varied variety of finished petrochemical products which is not possible from any single grade of crude oil. The main function of an oil refinery is to manufacture as economically as possible, the required quantities of petroleum products from the crude oil supplied to it.

To achieve these objective appropriate processes must be applied through necessary plant and equipment to get desired products.

Typically, 5 types of processes are used in general refining process:

- Separation processes
- Conversion processes
- Treating processes
- Blending processes
- Other processes

PHYSICAL SEPARATION:

In the initial phase of refining process, crude is separated into its major constituents using atmospheric distillation, vacuum distillation and gas processing. These processes separate the crude into common boiling point fractions.

The structure of hydrocarbon molecules is not changed and no new compounds are formed. The techniques used include:

DISTILLATION:

This method uses the property of boiling point to separate products according to molecular size. It is the first step to separate the crude oil into its main fractions and is the most important process in the refinery.

Fractional distillation is useful in separating a mixture of hydrocarbons with narrow differences in boiling points.

A typical distillation column with collecting trays is shown below:

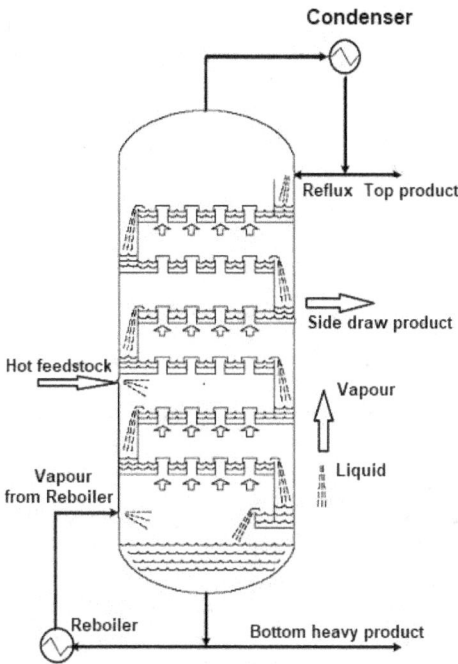

Figure 11: A typical distillation column

Distillation plays an important role in refining the products to meet marketing specifications. Crude distillation separates raw crude oil into number of intermediate refinery streams known as "crude fractions" or "cuts" which are characterized by their boiling points. Each fraction leaving the CDU (crude distillation unit) is defined by unique boiling point range and is made up of hundreds of hydrocarbon compounds, all of which have boiling points within the cut range. These fractions include light gases, naphtha, gas oil, and residual oil. Each fraction is then sent to a different refinery process for further processing.

The main feature of various products is the ability to vaporize according to the size of the molecules. Products like wax which solidify under normal conditions require heating to a high temperature before they liquefy and still higher temperature before they vaporize.

Volatility is related to the boiling point of a liquid. A liquid with low boiling point is more volatile than one with a higher boiling point.

The various fractions of crude oil are vaporized between 200- and 500-degrees Fahrenheit each at its own different temperature. The least dense products pass off first, then the lightest and so on. With gradual increase in heat, all the desired products are obtained.

Each product obtained in this Process is called a fraction; hence this process is called "Fractional distillation."

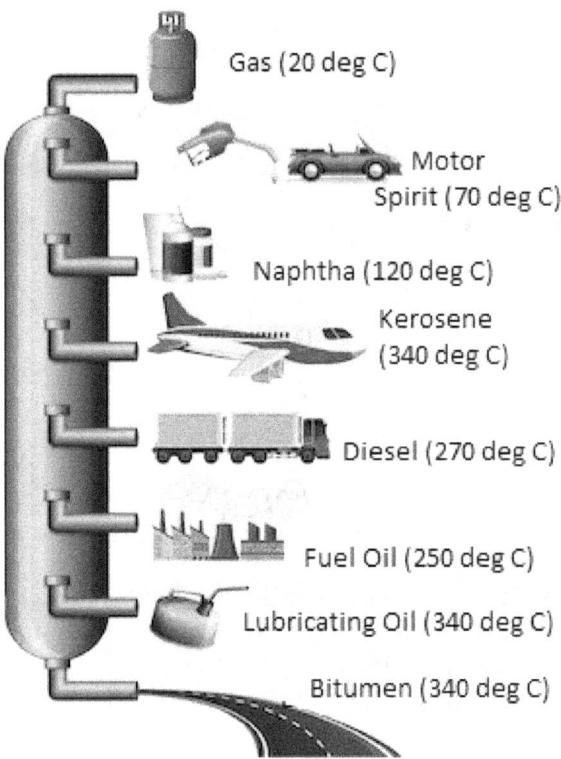

Figure 12: A typical Distillation

Distillation can be further classified as "Atmospheric distillation" and "Vacuum distillation".

ATMOSPHERIC DISTILLATION:

Consider atmospheric distillation as an industrial-sized version of a massive domestic pot. Crude oil is cooked to a high temperature and vaporized in this massive pot. The fumes climb through a towering column and cool as they do so. The vapors in the column are cooled to the point where they condense back into liquid at various heights. The crude oil is divided into different components by this amazing column, which functions like a massive distillation puzzle. Heavy substances like diesel and leftover oil sink to the bottom, while lighter substances like gasses and naphtha ascend to the top.

The main fractions or "Cuts" have specific boiling point range and are classified (based on volatility) as gases, light distillates, middle distillates, gasoil and residue.

The desalted crude feedstock is pre-heated and sent to direct fired crude charge heater and then goes into vertical distillation column just above the bottom. A temperature in the range of 330 to 375 deg. Centigrade and pressure slightly above atmospheric pressure is maintained to avoid thermal cracking. Fractions other than heavy fractions flash into vapor. As the hot vapor travels higher & higher in the column, its temperature is reduced. Heavy fraction (heavy fuel oil & asphalt) residue settles at the bottom of the column.

The other major products like gasoline, kerosene, heating oil, uncondensed gases etc. are drawn off at successive higher points of the tower.

Reflux is provided by condensing the overhead vapors of the tower and returning a portion of the liquid to the top of the tower. To obtain maximum reflux the liquid is removed from the tower, cooled by heat exchanger and returned to the tower. Also, a part of the cooled side-stream is returned to the tower. The cold stream condenses more vapors coming up and increases the reflux.

Normally, crude towers do not use reboilers, several trays are put below the flash zone and steam is passed below the bottom tray to strip any remaining gas oil from the liquid in flash zone, resulting in high flash point bottoms. The atmospheric fractionator contains 40 to 50 fractionation trays. Generally, 6 to 8 trays are needed for each side-stream product.

The liquid side-stream withdrawn from the tower contains low boiling point components which are stripped from each side-stream in small stripping tower containing 5 to 10 trays with stream introduced under the bottom tray. The light ends and steam are vented back into the vapor zone of the fractionator above the corresponding side-draw tray.

Pentane and heavier fractions of the vapors passing out from top of the tower are condensed by overhead condenser of the atmospheric tower. A portion of this condensate is returned to the top of the tower as reflux, and the remainder is sent to stabilization section of the refinery where butane and propane are separated from the gasoline.

A simplified version of Crude distillation unit is shown below:

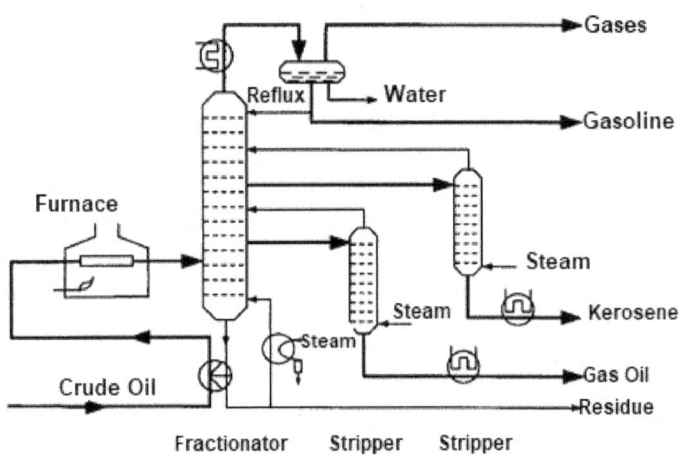

Figure 13: Crude distillation unit (simplified)

During atmospheric distillation the bottom column contains high boiling point hydrocarbons which will have to be separated into its constituents by vacuum distillation. After heating the crude oil in the still for several hours, it gets vaporized; the vapors are passed through pipe going from still to cooling tank which is filled with cold water.

As the vapors pass through hundreds of feet of pipe in the cooling tank, the major portion of vapors change to liquid. This liquid flow to tail house is monitored by "Stillman". The Stillman observes the liquid flowing through the "Lookbox" (also called receiving house) and takes sample from time to time to find the density (specific-gravity).

Since the standard range of specific gravity of the products under the distillation process is known along with the color and "feel" (acquired by long years of experience of Stillman) he can tell when complete first fraction has passed over. The receiving storage tanks are changed for receiving different products under advice from "Stillman" based on above testing criteria.

As can be envisaged there are chances of human error in above process of visual checking for color and specific gravity to make exact "Cut" from one fraction to another, hence most of the fractions are distilled again to get fine Cut.

This is a highly technical issue and is done with care, which is why there are more than 300 types of fractions available from some types of crude oil. The temperature of the crude oil inside the still is measured by Pyrometer (a kind of thermometer). During distillation the distillates are received in the following sequence:-

Sequence of Products obtained	Temperature range (degree Fahrenheit)	Products obtained
Initial vapors		Gas, Crude Naphtha
Crude Naphtha is redistilled	200 to 250	Naphtha, Benzene, Gasoline
	250 to 300	Kerosene
	300 to 350	Gas Oil
	350 to 500	Paraffin distillate
Residue is Cylinder stock		From Cylinder stock we obtain Asphalt, Coke, Road oil, Flux

Gas and Crude Naphtha are received as first distillate, which is redistilled to get Naphtha, Benzene and Gasoline and other products shown in table above. Naphtha is gasoline boiling range material which is sent to upgrading units for octane improvement and sulfur reduction, and later sent for gasoline blending.

The distillates are further treated and blended to aviation fuel, diesel and heating oil. The residual oil (or bottoms) is sent to other conversion units or blended to heavy industrial fuel & asphalt. These are low value products and most refineries try to upgrade them into more valuable lighter products.

VACUUM DISTILLATION:

In the process of atmospheric distillation of heavy fractions of crude oil, the furnace outlet temperatures are very high which could result in thermal cracking resulting in loss of product and equipment fouling.

The heavier fractions left over from the atmospheric distillation process are subjected to another round of processing known as vacuum distillation. This phase entails lowering the pressure to allow the remaining heavy components to be separated without the use of extremely high temperatures.

The process of vacuum distillation is the same as atmospheric distillation except the fact that larger diameter columns are used to maintain comparable vapor velocities at reduced pressures. In vacuum towers instead of trays we use random packing. Distillation is carried out with absolute pressure of 25 to 45 mm of mercury in the tower flash zone area.

First phase vacuum tower may produce gas oil, base oil for lubricant base stock and heavy residue oil for propane DE asphalting. Second phase tower operates at lower vacuum and distills surplus residue from atmospheric tower which is not used for lube stock processing. Ideally the vacuum towers help in separating catalytic cracking feedstock from surplus residue.

Steam is added to the furnace inlet to increase the furnace tube velocity and minimize coke formation in the furnace. Steam also decreases the total hydrocarbon partial pressure in the vacuum tower. Vacuum towers up to 50 feet in diameter are used in some refineries.

CRUDE FRACTION TO FINAL PRODUCT

Fraction	Boiling Range (deg Centigrade) approx.	Next stage	Final Product
LPG	-40 to 0	Sweetener	Propane
Light Naphtha	39 to 85	Hydrotreater	Gasoline
Heavy Naphtha	85 to 200	Catalytic Reformer	Gasoline, Aromatics
Kerosene	170 to 270	Hydrotreater	Aviation fuel, Diesel (grade-1)
Gas Oil	180 to 340	Hydrotreater	Heating oil, Diesel (grade-2)
Vacuum Gas Oil (VGO)	340 to 566	FCC Hydrotreater Lube plant Hydrocracker	Gasoline, LCO, Gases Fuel oil, FCC feed Lube base stock Gasoline, Jet fuel, Diesel, FCC feed, Lube base stock
Vacuum Residue	>540	Coker Visbreaker Asphalt unit Hydrotreater	Coke, Coker gas oil, Residue, Visbreaker gasoil, Deasphalted oil, Asphalt FCC feed

"It is interesting to note that the refinery capacity is expressed in terms of crude oil distillation throughput capacity (in million metric tons per annum or **MMTPA**)."

CRYSTALLIZATION:

This method uses the property of melting point, solubility, filtration or centrifuging. The high molecular weight paraffin hydrocarbons essentially give paraffin wax which readily separates by crystallization when an oil fraction containing wax is cooled. Also, the wax content of lubricating oils needs to be removed by this method as the lubricating oil when used in engines, will congeal at low temperatures which is undesirable.

ABSORPTION:

This method uses the property of solubility in a liquid to separate light gases from heavier ones. Natural gas contains low boiling point hydrocarbons. Amongst these hydrocarbons propane and butane can be liquefied under moderate pressure and atmospheric temperature, whereas hydrogen, methane and ethane can be liquefied only under high pressure and at low temperature.

Ethylene and propylene are produced in cracking units and are also easily liquefied. These hydrocarbons are valuable as **LPG** or as feed stock for the chemical industry. The operation of an absorption process is similar to that of a distillation process. The wet gas which contains heavy components is introduced into the bottom of the absorption column and lean absorption oil towards the top. The unabsorbed gas i.e., the dry gas leaves the top of the absorber while the fat absorption oil leaves the bottom.

SOLVENT EXTRACTION:

This method uses the property of miscibility with another component to separate Paraffins from Aromatics, hence the process is also called as "aromatic extraction". This process uses a solvent in which one group of oil components usually the aromatics is dissolved.

If the unrefined oil is thoroughly mixed with the solvent and then allowed to settle two layers are formed. The lower phase will contain the dissolved components for example the aromatics and the upper will contain undissolved components like paraffin.

The unrefined oil is in this way is separated into a raffinate containing the less soluble components and the extract containing the most soluble components After separation of these phases the solvent is recovered by distillation from both phases and reused in the process.

This process delivers good quality kerosene for lamps and jet engines. Also, benzene, toluene and xylenes are obtained by this process which is primary feedstock for petrochemicals. The extract has high aromatic content and is suitable as a tractor fuel or as feedstock for chemical products.

CHEMICAL CONVERSION

This method uses cracking (thermal & catalytic), polymerization, alkylation, dehydrogenation and/or isomerization.

C/H RATIO:

This ratio is of major importance in any refinery and is basically Carbon to Hydrogen ratio. In any crude oil, the heavier the boiling range fraction, the higher is its C/H ratio. This is also true of refined products, the heavier the refined product, higher is its C/H ratio.

Hence it is important for the refineries to reduce the C/H ratio for getter better economic value products. The C/H ratio can be reduced in 2 ways:

- By reducing the carbon atoms
- By increasing the hydrogen atoms

The reduction in carbon atoms is brought about by rejecting carbon atoms in the form of petroleum coke, or by adding hydrogen. FCC and Coking processes are used for rejecting excess carbon atoms, while Hydrocracking is used for removing hydrogen.

Chapter Eleven

CATALYTIC CRACKING AND HYDROCRACKING:

This process involves chemical reactions which fracture (crack) the large, high boiling point range hydrocarbons (these are of low economic value) into lighter molecules which after blending are suitable for blending to gasoline, diesel, petrochemical feed stocks and other high value lighter products.

The ignition quality of fuels depends on the molecular configuration of the hydrocarbons. With the development of internal combustion engine, it became necessary to produce lighter fractions from crude oil. For spark ignition engines Naphthenes and aromatics are better than the straight chain compounds where is for diesel engines which require compression straight chain compounds give better ignition properties.

THERMAL CRACKING:

Thermal cracking is the process of subjecting hydrocarbons to relatively high temperatures over a prolonged period of time during which the larger molecules of heavier oils are broken down into smaller ones like those needed for gasoline and gas oil.

The heavy petroleum distillates are then subjected to temperatures from 425 degrees centigrade to 500 degrees centigrade at pressures between 28 to 355 pounds per square inch. The long chain paraffinic hydrocarbon molecules are broken down into number of smaller ones by rupture of the carbon-carbon bonds.

The final products consist of light hydrocarbons in the gasoline and gasoil range. Thermal cracking improves the quality and quantity of gasoline as compared to the straight distillation process which was used earlier.

CATALYTIC CRACKING:

The feedstock used in Catalytic cracking is usually gas oil, from distillation, coking and DE asphalting processes. In this process the cracking is carried out using heat and pressure in the presence of catalyst to produce better quality of fuel. The feedstock typically has boiling range of 350 to 550 deg C

A catalyst changes the rate of chemical reaction, and remains chemically unchanged at the end of the reaction. Many substances are used as catalysts, most common being synthetic materials based on Silica and Alumina. They are used as pellets or as fine powder.

Catalysts increase the rate of reaction and also improve anti-knock quality of gasoline. Thermal cracked gasoline gives anti-knock value of 60 to 70 only, whereas catalytically cracked gasoline gives anti-knock value up to 93. Of late 2 stage catalytic cracking has been developed which gives better yield of gasoline and less coke.

FLUID CATALYTIC CRACKING (FCC):

FCC is one of the most important refining processes in the crude oil distillation process. FCC requires high temperature, low pressure and catalyst to convert heavy gas oil to light gases (C1 to C4 and olefins), feedstock for petrochemicals, blend-stock for gasoline (FCC naphtha) and diesel fuel blend-stock (light cycle oil).

FCC offers following advantages:
a) High yield of gasoline and distillates
b) Operating flexibility to use various types of crudes based on availability and economics
c) Low operating cost
d) High reliability

In general, the FCC unit accounts for more than 40 % output of gasoline and diesel fuel. The ratio of gasoline to diesel produced also known as G/D ratio is also controlled based on demand pattern varying by climate and country. For example, in US the G/D ratio is higher in summer than in winter based on fuel demand pattern.

Sulfur is deterrent to FCC catalysts and must be removed from FCC feedstock. To take care of this issue many refineries have desulfurizing units before FCC unit and the feedstock for FCC is routed first through these units to remove/reduce sulfur content.

The unreacted feed from FCC unit is called "Slurry oil" and is sent to Coking unit in the refinery.

HYDROCRACKING:

Hydrocracking converts distillates, gas oil and other heavy streams to gasoline and distillates. It is a catalytic process which operates at moderate temperature and high pressure.

In this type of catalytic cracking, a small amount of hydrogen is introduced during the process to crack distillates. This process is carried out at temperatures between 250 and 430 deg C and at pressure between 500 to 3000 psi.

Hydrocracking gives high yields of light products and operating at one extreme it can convert all of its feed to gasoline blend stock with yields as much as 100 % on feedstock. Also, it is possible to get jet fuel and diesel with combined yield of up to 90 % by volume along with small amounts of gasoline.

This process has marked advantage over FCC; the introduction of hydrogen to hydrocracker unit not only leads to cracking, but also other reactions which remove sulfur from hydrocracked streams. The hydrocracked streams are not only low in sulfur but also low in aromatic content.

Aromatics in the distillate give poor engine performance and poor emission characteristics in diesel. Due to chemical reactions in the hydrocracking, the aromatic rings are broken up and produce premium blend stock with excellent performance and emission characteristics.

Very good quality of lubricating base oils is obtained by hydrocracking of vacuum gas oil. In this process any fraction of asphalt or naphtha can be processed to get desired product.

Hydrocracking produces low sulfur products, better than FCC or Coking, but is expensive to build and operate mainly due to high hydrogen consumption.

VISBREAKING:

In order to improve certain properties, a procedure called Visbreaking is also used. Visbreaking is the lowering of the viscosity of very viscous oils by mild thermal treatment making them suitable for further handling and increasing the value and improve the ignition quality.

Nowadays Thermal cracking has been replaced by Catalytic cracking which produces gasoline of higher-octane number.

COKING:

Coking converts low value residual fuel into high value petroleum coke and gas oil and light petroleum stocks. Coking is a thermal, non-catalytic conversion process. Coking is the primary means of converting residual oil to valuable lighter products.

CATALYTIC CRACKING AND HYDROCRACKING: 115

The cracked products from coking are as follows:
- Coker naphtha (low quality naphtha)
- Coker distillate (distillate stream)
- Coker gas oil
- Petroleum coke

The Coker gas oil can be used as additional FCC feed, but it is less valued as FCC feed because it contains high levels of sulfur.

The petroleum coke produced in Coker unit can be used as fuel in refinery or sold to power plants.

DELAYED COKING:

This process upgrades low value residual stocks, heavy vacuum residue and asphalt into gas, gasoline, middle distillates and coke. This is one of the oldest petroleum conversion processes and it is used for increasing diesel fuel and producing low Sulfur petroleum coke.

High Sulfur stocks are first desulfurized to be used for production of high-quality low Sulfur petroleum coke. On calcination this gives low Sulfur calcined coke which is required mainly by the metallurgical industries.

REFORMING:

In this process straight run gasoline is cracked under strict controlled conditions so that conversion is limited. This is achieved thermally or by the aid of a catalyst.

Reformers carry out catalytic reactions on naphtha streams and increase octane of these streams by as much as 50 octane numbers. The reformer output is called reformate and is high octane gasoline blend stock.

The primary chemical reactions in reforming produce aromatics in the gasoline boiling range having high octane rating.

THERMAL REFORMING:

Thermal reforming is done at high temperatures requiring larger reaction times than thermal cracking. In order to prevent reaction from proceeding too far, resulting in excessive gas formation and lower gasoline yield, the hot products leaving the furnace are cooled immediately by quenching with cold oil.

Reforming is usually carried out at temperatures around 550 degrees Centigrade & pressure ranging from 80-100 kg/cm^2.

CATALYTIC REFORMING:

Thermal reforming gives low yield of ignition quality gasoline and has been replaced now days with catalytic reforming. Catalytic reforming is core reforming process which uses a catalyst consisting of alumina carrier containing 0.4 to 0.75 % platinum. Typically, small amounts of Chlorine and Fluorine are also present as activators.

Straight run gasoline is introduced with a surplus of hydrogen in the process. The normal reactions of thermal cracking take place, but due to the presence of catalyst, isomerization and cyclisation also take place. The resulting product is rich in aromatics and iso Paraffins. The low ignition quality components are converted to high ignition properties.

This process is used to premier gasoline and aviation gasoline (jet fuel). Reforming can produce reformates with octanes > 100 RON.

Reformers produce hydrogen as co-product. The aromatic grades needed by chemical industry are also produced by this process.

PROPANE DEASPHALTING:

DE asphalting utilizes the property of solubility of hydrocarbons in propane. Butane, pentane and propane are used as solvent, out of which liquid propane is mostly used. Vacuum residue is fed through a countercurrent DE asphalting tower. The alkanes dissolve in propane, whereas aromatic compounds (asphaltic materials) do not dissolve. Asphalt is then sent for thermal processing.

Asphaltic bitumen is obtained by the distillation of asphaltic base crude oils. Bitumen is manufactured in various grades from soft to hard varieties. Harder grades of bitumen are obtained by the removal of volatile material during the distillation process.

TREATING PROCESSES

The products obtained by separation and conversion processes are further treated to bring them up to marketing specifications in respect of color, odor and stability etc. The removal of undesirable components (sulfur, nitrogen, heavy metals) from crude oil fractions is carried out by treating processes.

If treated with hydrogen it is known as hydro-treating, otherwise it is chemically treated by the use of chemicals such as sulfuric acid. Further treating is also needed to remove the chemicals like sulfuric acid if used in the treating process.

The main purpose of treating is to:

- Protect the catalyst from deactivation, resulting from prolonged contact with sulfur, nitrogen & heavy metals
- Meeting the desired marketing specifications of refined products

The most widely used technology to achieve the above objective is catalytic hydrogenation or hydro treating.

HYDROTREATING:

In this process sulfur, nitrogen and heavy metal compounds are removed by converting them to hydrogen compounds in the presence of a catalyst.

The process is called hydro-desulphurization when Sulfur is removed in the form of hydrogen sulphide. Similarly, when nitrogen is removed the process is called hydro-DE nitrification. The catalyst used is a mixture of cobalt and molybdenum oxides supported on aluminum alumina carrier.

It also converts aromatics in the kerosene into Naphthenes and improves the smoke point.

HYDRO-REFINING:

When hydro treating is carried out at high temperature, high pressure and hydrogen concentration, the process is called hydro-refining. For lubricating oils hydro treating is done at low severity to modify certain characteristics to meet market specifications. Mild hydro treating is commonly called hydro-finishing.

In general, the refineries have many hydro treating units operating on different crude fractions, feed stocks and blend stocks (varying from light naphtha to heavy residue). These serve many purposes:

a. The catalytic reformers have naphtha Hydrotreater which reduce the sulfur content of reformer feed to less than 1ppm, in order to protect the reformer catalyst.

b. Some reformers remove benzene from reformate by the help of post-Hydrotreater (also known as benzene saturation units).

c. Some FCC units in refineries which are processing sour crude or producing low-sulfur gasoline and diesel, have FCC feed Hydrotreater. These reduce the emission of sulfur oxides and protect the catalyst from poisoning by nitrogen and other heavy metals. They also improve cracking yield and reduce sulfur content of FCC products.

d. Sulfur from mixture of blend stocks is removed by distillate Hydrotreater. They also help in improving Cetane number specification for diesel.

CRUDE DESALTING:

Crude oil contains water, salt and suspended solids which are not needed in final product. These contaminants need to be removed to reduce corrosion, equipment fouling and to prevent poisoning of catalysts in processing units.

During Distillation of crude oils, the process generates naphthenic acid, hydrochloric acid and sulfur compounds which are harmful to the costly crude distillation unit commonly known as "CDU". In order to reduce/eliminate these unwanted compounds the most common method used is injection of caustic soda in the crude stream.

In chemical desalting, water and chemical surfactants are added to the crude and heated so that salt and other impurities either dissolve or get attached to water and are removed after settling in the tank.

The best method for reducing the salt content used nowadays is "Electrical/Electrostatic Desalting" operation in which hot water acts as extraction agent. In this process the crude oil is preheated to temperatures between 120 degrees centigrade to 130 degrees centigrade and then charged to desalting vessel. Preheating reduces viscosity and surface tension which helps in easy mixing and separation of water & an emulsion of oil and water is obtained.

Desalting process is of 2 types: single stage desalting and 2-stage desalting (explained diagrammatically below):

CATALYTIC CRACKING AND HYDROCRACKING: 121

SINGLE STAGE ELECTROSTATIC DESALTING PROCESS

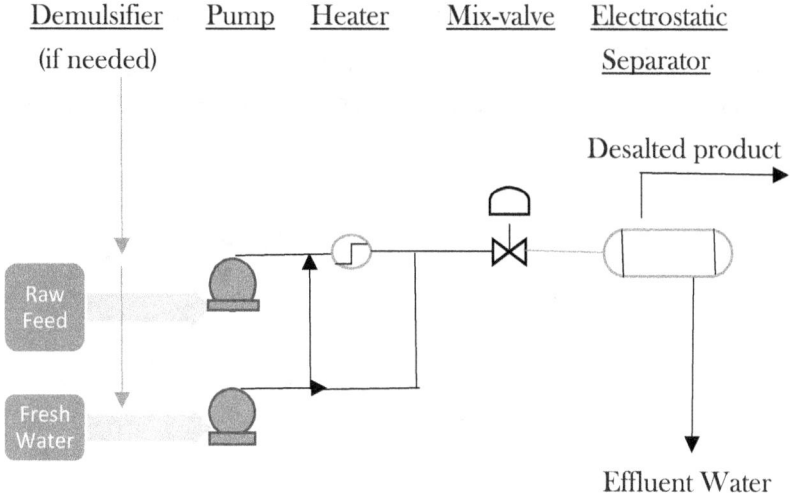

Figure 14: Single stage Electrostatic Desalting

Emulsified crude oil is subjected to high voltage of 20,000 to 25,000 volts (either AC or DC) electrical field in the desalter where the water coalesces and separates from the emulsion together with the salt from the crude. Salted crude overflows from the water and along with the salt and crude sludge remains at the bottom of the desalter vessel.

TWO STAGE ELECTROSTATIC DESALTING PROCESS

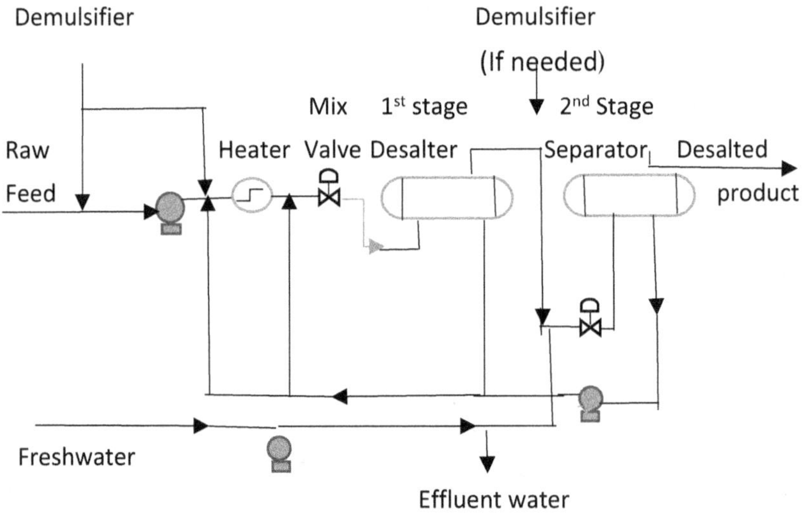

Figure 15: Two stage Electrostatic Desalting

In this operation salt content of the crude is brought down to the extent of 95% in single stage desalting, whereas in 2-stage desalting 99% efficiency can be obtained. Surfactants are added only when crude oil has a large quantity of suspended solids.

Emulsions are formed when pH of brine exceeds 7. It is desirable to keep pH in the range of 6 to 8 for better dehydration in electric desalter. The pH is controlled by adding water, caustic or acid to inlet or recycled water.

Waste water and contaminants are discharged from the bottom of the settling tank to waste water treatment facility or oil water separator.

SULFURIC ACID TREATMENT:

During earlier days sulfuric acid treatment was used for the removal of sulfur, aromatics and olefins. The process was applicable to treatment of straight run naphtha.

The major drawback of this process was that appreciable quantity of material was lost due to the formation of acid sludge and there were problems in disposal of sludge. Later processes like hydro finishing were developed which eliminated the drawbacks of acid treating and improved the product quality.

CLAY REFINING:

This process is used to improve the color and stability of the acid treated oils and given as a final treatment. During the treatment of lubricating oils and waxes the slurry of oil and clay is heated to temperatures of 120 degrees centigrade to 150 degrees centigrade.

The clay used in the process is generally activated. This process is not used for Lube refining but it has been replaced by hydro finishing where the yield is of the order of 99% by weight.

CAUSTIC WASHING:

In this process alkaline reagents like calcium, ammonium and sodium hydroxide are used for the removal of hydrogen sulfide during the refining process of crude.

SWEETENING:

During gasoline and kerosene production, low boiling point compounds called Mercaptan tend to accumulate. This mercaptan impart foul odor and are highly corrosive. They also reduce the octane rating of gasoline. Therefore, it is important to remove or convert them into less harmful form.

ASPHALT BLOWING:

This process polymerizes the asphaltic residue by oxidation which results in increasing their melting temperature and hardness. Heated air is blown through a heated batch mixture of oil in a continuous process. Heat is generated in the process (reaction is exothermic), and temperature is controlled through quench steam. To increase the rate of reaction, ferric chloride or phosphorus pentoxide is sometimes used as catalyst.

DOCTOR TREATMENT:

Doctor solution consists of an aqueous mix of caustic soda and sodium plumbite. This is used during the process to remove mercaptan which is converted to lead mercaptide.

HYPOCHLORITE TREATMENT:

This process is used for sweetening Naphtha using Sodium and Calcium hypochlorite.

MEROX PROCESS:

In this process Petroleum distillates are chemically treated to remove mercaptan which is converted into disulfides. The total Sulfur content remains unchanged.

There are 2 types of Merox sweetening: in the first type which is solid bed sweetening the Merox catalyst is held on a solid support of activated charcoal. In the second type which is called liquid sweetening, hydrocarbon air and caustic solution containing the dissolved Merox catalyst is used for sweetening.

AFTER TREATMENT:

In most of the cases no after treatment is required. In case of jet fuels or aviation turbine fuels, where quality is of paramount importance, to meet the rigorous thermal stability and water tolerance test the treated stream is given auto wash followed by sand bed coalescers or salt dryer for removing the entrained water.

OTHER PROCESSES

In order to meet the marketing specifications, the refinery grade products obtained by the refining processes need further processing, the most important of which are blending emulsifying and grease making.

BLENDING:

All refineries, regardless of their size need product blending facility to meet the product quality requirement and applicable government standards. These standards may specify some of the below parameters:

- Density
- Boiling range
- Volatility
- Sulfur content
- Aromatic content
- Octane number
- Cetane number
- Smoke point
- Freezing point

Some of the products obtained by above methods may be suitable for marketing directly, but more often these need to be modified by blending with other compounds to make them fit for usage in various applications. Common example is different grades of bitumen, wax, automotive lubricants and greases.

Paraffin wax is a solid substance when pure and at ordinary temperatures is soluble in the crude oil. The proportion of gaseous and solid constituents is variable from one type of oil to another, and even in oils from different wells drawn from the same place.

The blending of gasoline is very complex and highly automated operation. The blend properties are continuously monitored by online analysers achieved through complex computer controls. Gasoline is a mixture of 5 to 10 blend stocks, whereas diesel is a mixture of 4 to 6 blend stocks.

BLENDSTOCK CHARACTERISTICS:

Typical values for gasoline & diesel blend stocks are tabulated below (figures are for unprocessed raw streams, which will be processed further to improve their properties):

Blend stock	Octane RON	Sulfur (ppm)	RVP (psi)	Aromatics (vol%)	Benzene (vol%)	Olefin (vol%)
Straight run Naphtha	70 - 71	110-120	12	-	-	-
Isomerate	80 - 82	1	13	-	-	-
Reformate	95 - 97	< 4	5	60	5	-
Alkylate	92 - 94	< 10	3	-	-	-
FCC Naphtha	90 - 92	500 – 1500	5	25	1	30
Coker Naphtha	85 - 88	500 +/- 10	19	0.5	0.5	50
Hydrocracked Naphtha	75 - 78	< 4	11	2	2	-

Blend stock	Sulfur (ppm) approx.	Cetane number	Aromatics (vol %)	Specific gravity
Straight Run Kerosene	3000	45	19	0.82
Straight Run Distillate	7000	53	21	0.85
FCC Light Cycle Oil	12500	22	80	0.93
Coker Distillate	32000	33	40	0.89
Hydrocracked Distillate	100	45	20	0.86

Figure 16: Blend stock Characteristics

EMULSIFYING:

Certain petroleum products are marketed as emulsions with water. Bitumen emulsions and marine lubricants are the best examples of oil-water and water-oil emulsions and both are made with petroleum wax. Oil and water get emulsified at normal ambient temperature but wax and bitumen must be heated to make them liquid and the water phase also has to be heated to maintain the fluidity during the process.

GREASE MAKING:

Grease is prepared by charging oil into an autoclave along with alkali and some fatty material. The mixture is heated at pre-defined temperature and stirred until soap formation is complete. Next it is cooled with stirring until there is formation of grease like structure.

Inorganic thickeners may also be used in particular types of greases, and the process varies with the type of thickener used. The Lithium based greases are manufactured in special continuous process.

SULFUR RECOVERY:

Refinery gas streams contain hydrogen sulphide gas (H_2S) from which commercially viable elemental Sulfur is recovered by the Sulfur recovery unit.

BLOWDOWN SYSTEM:

This system helps in the safe disposal of vapor & liquid hydrocarbons discharged from pressure relief devices in the refinery. The blow down material is collected through a series of drums and condensers and collects as liquid or vapor. The liquid is recycled whereas the vapors can be flared off or recycled.

PROCESS HEATERS:

Process heaters are furnaces which refineries use for raising the temperature of feed stock up to level needed to carry out the reactions.

FUGITIVE EMISSIONS:

Sources of fugitive emissions may include hydrocarbon vapors from process equipment, valves, flanges, pumps, cooling towers, oil-water separators, transfer equipment, storage, drains, leaks and spills etc.

UTILITIES AND SUPPORT OPERATIONS:

Support operations like control of emissions to air, water and other important functions are also of great significance, some of them being:

- Production and recovery of hydrogen
- Sulfur recovery
- Waste water treatment with the help of oil water separator
- Oil movement and storage
- Electricity & steam generator
- Fire hydrant system
- Adequate Firefighting and safety systems

In any refinery, substantial amount of hydrogen is needed as input for hydrocrackers and Hydrotreater. Partially the requirement is met through hydrogen produced by reformer, and additional requirement needs to be purchased from outside. Due to high cost of procuring hydrogen from outside, most refineries have facilities for recovery and recycling of spent hydrogen in hydrocracking and hydro treating streams.

The refinery process streams use fuel and steam to provide energy for driving chemical reactions. Electricity is needed to run the pumps, motors and compressors which can be purchased or produced locally if the refineries have their own generation facilities (steam boilers & power generation facilities).

Chapter Twelve

DESULFURIZATION & ENVIRONMENTAL COMPLIANCE

Within the oil refining industry, desulfurization is a critical step that is closely related to environmental regulations. The petroleum industry is placing a greater emphasis on lowering the sulfur content of gasoline as consumer demand for sustainable practices and cleaner fuels rises. This thorough investigation breaks down the mystery surrounding desulfurization, reveals how important it is to maintaining environmental compliance, and explores the laws and rules that have shaped this revolutionary path in the petroleum industry.

Understanding Desulfurization

THE SULFUR PROBLEM:

Sulfur, a naturally occurring ingredient in crude oil, creates environmental risks when used in fuels. During burning, it emits sulfur oxides (SO_x), which contribute to air pollution, acid rain, and negative health impacts. Desulfurization, as a result, becomes an important step in refining, with the goal of removing or decreasing sulfur compounds to meet severe environmental criteria.

DESULFURIZATION TECHNIQUES:

1. **Hydrodesulphurization (HDS):**

    Sulfur molecules are selectively eliminated in the hydrodesulphurization process by interacting with hydrogen in the presence of catalysts. As a result, sulfur-containing molecules are converted into hydrogen sulfide, a less toxic chemical.

2. Catalytic Cracking:

Catalytic cracking is the process of converting big hydrocarbons, such as sulfur compounds, into smaller and more valuable products. This technique helps with desulfurization as well as the manufacturing of high-demand fuels like gasoline.

3. Oxidative Desulfurization:

This new technology uses oxidative reactions to remove sulfur from fuels selectively. It is especially excellent at removing refractory sulfur compounds that are difficult to remove using traditional procedures.

Significance of Desulfurization in the Petroleum Sector

ENVIRONMENTAL IMPACT:

1. Air Quality Improvement:

Decreasing the sulfur concentration in fuels results in decreased sulfur oxide emissions, contributing to improved air quality and reducing the environmental effect of combustion.

2. Health Benefits:

Lower sulfur emissions mean lower health risks for the people, particularly in metropolitan areas where air pollution is a major concern.

GLOBAL REGULATORY LANDSCAPE:

1. International Maritime Organization (IMO):

International Maritime Organization (IMO) regulations, such as the International Convention for the Prevention of Pollution from Ships (MARPOL Annex VI), establish global sulfur restrictions for marine fuels. The IMO's global sulfur cap of 0.50% for marine fuels, which has been in effect since 2020, aims to reduce sulfur emissions from shipping.

2. **European Union (EU):**

   The EU has imposed rigorous sulfur content standards in fuels for both road and maritime use. Euro 6 emission rules, for example, require automobiles to use ultra-low sulfur diesel.

United States Environmental Protection Agency (EPA):

   To limit emissions from automobiles and non-road equipment, the EPA has established sulfur content guidelines for gasoline and diesel fuels. Tier 3 gasoline rules attempt to reduce sulfur levels, resulting in cleaner car emissions.

Challenges in Desulfurization

TECHNOLOGICAL CHALLENGES:

1. **Cost Implications:**

   Implementing modern desulfurization technology frequently necessitates significant capital expenditure as well as operational costs, which has an impact on the overall economics of refining.

2. **Catalyst Deactivation:**

   Desulfurization catalysts can deactivate over time, needing constant regeneration or replacement, increasing the complexity and cost of the refining process.

ENERGY EFFICIENCY:

1. ### Hydrogen Consumption:

   Hydrodesulphurization, which relies on hydrogen, necessitates a large volume of this fuel. The energy-intensive nature of hydrogen production raises sustainability and efficiency issues.

Innovations and Future Trends

EMERGING TECHNOLOGIES:

1. **Ionic Liquids:** Ionic liquids have the potential to selectively remove sulphur compounds from fuels. Their distinct features make them a viable alternative to existing desulfurization processes.
2. **Bio desulfurization:** Biological techniques that use enzymes or microorganisms provide environmentally friendly desulfurization options. Methods of bio desulfurization target specific sulfur compounds while reducing environmental impact.

HYBRID APPROACHES:

1. **Integrated Refining Processes:** Refineries are looking into integrated techniques in which desulfurization is integrated into other refining processes. This all-encompassing approach improves overall efficiency while lowering environmental impact.
2. **Advanced Catalysts:** On-going research focuses on the development of improved catalysts with better activity, selectivity, and resistance to deactivation, solving some of the issues associated with traditional desulfurization processes.

Chapter – Thirteen

PETROCHEMICAL INTEGRATION AND DOWNSTREAM PRODUCTS

Transportation of crude in oil business starts from wells and ends at the refinery. The various equipment used in the process are:

- Pipelines
- Pumping stations
- Storage tanks
- Ships

Main trunk pipelines vary from 6 to 24 inches in general, with operation pressure up to 550 psi. They are made up of high-grade seamless steel and tested to internal pressure of 1000 psi.

To keep the crude oil moving at desired rate there are Booster or pumping stations along the pipeline route. Some oils are so heavy that they have to be heated before putting into the pipeline.

The pumps are so powerful that through an 8-inch pipeline they are powerful enough to deliver 30,000 barrels of medium light oil in 24 hours at a line pressure of 800 psi.

When there is a change in grade of oil to be inserted in pipeline, there is a need to put header in between different grades. A water slug of 3 feet long is put in between the grades to separate the grades. After the pumping is completed, another header is inserted and then original grade is inserted.

The pipelines are generally owned by Refineries, so their storage tanks are usually built near the refineries in groups called "Tank Farms." They cover hundreds of acres and may contain up-to hundreds of tanks. The Tank sizes vary from 37500 to 55000 barrels, with diameter ranging up to 115 feet and height up to 35 feet. The largest crude oil tank in the world is having a capacity of 250,000 Cubic meters.

"Elizabeth Watts" was the first Vessel (ship) of 224 MT capacity to transport Crude oil between Philadelphia and London in Nov. 1861. The first cargo consisted of 1329 barrels of oil. It took 10 days for loading and 12 days for unloading. Zoroaster was the first modern type of tanker built in Russia by Ludwig Nobel in 1878, after which the tanker building industry progressed rapidly.

In the earlier years of the Petroleum Industry, "Coal Oil" was the only product obtained from Crude oil. The remaining residue which was around 91% contained wax, lubricating oil, and gasoline and was thrown away as useless waste or sold at cheap prices.

Chapter Fourteen

MEASUREMENT OF OIL & GAS

The measurement units used in Oil & Gas Industry are given below:

- **Crude Oil**: E & P companies use barrels (bbl.) per day as the unit of measurement of their output.

    1 barrel (bbl.) = 42 US gallons
    1 US gallon   = 3.785 liters
    1 barrel (bbl.) = 0.158987 Cubic meters

- **Natural Gas**: The natural gas output is measured in Cubic feet, for example Million cubic feet (Mmcf) or Trillion Cubic feet (Tcf)

- **Refined Petroleum fuel**: The refined liquid fuels are measured in Liters, Kiloliters, or Metric Tons (MT). For trading and intercompany transactions, the volume of oil is calculated in CuM at 15 degrees Centigrade and/or in MT.

1 KL = 1000 Liters

1 MT = 1000 Kilogram

The weight to volume conversion for liquid fuels is not by any direct formula because of the variation in volume at different temperatures. The liquid fuels expand with a rise in temperature, and contract with a fall in temperature. The density of the product plays a major role in the determination of weight/volume of a particular fuel.

For the information of reader's, a rough idea for a few petroleum products is given below (figs will vary with density/temperature).

1 Metric Ton of fuel = Density of Fuel at 15 deg C x 1000

For the information of readers, a rough idea on weight-volume relation of fuels is as under

1 MT of Naphtha = 650 to 700 liters
1 MT of Gasoline or Petrol = 700 to 750 liters
1 MT of Kerosene = 750 to 800 liters
1 MT of Gas Oil or Diesel = 800 to 850 liters

The above calculation is shown based on density range of these products at 15 deg. Centigrade which is given below:

Product	Density in grams/cc
Naphtha	= 0.650 to 0.700
Gasoline or Petrol	= 0.700 to 0.750
Kerosene	= 0.750 to 0.800
Gas Oil or Diesel	= 0.800 to 0.850

Chapter – Fifteen

CRUDE OIL PRICING

Crude oil price is of most important value on international commodity markets because of its importance. Emerging markets like China, India and Latin America greatly influence the price of oil, since they require oil to support their economic growth which results in rising energy consumption.

Crude oil supply is limited, and cannot be increased further, as finding and developing new oil reserves in recent years has become extremely difficult.

Crude oil is the most important energy source and the price of oil therefore plays an important role in industrial and economic development of any country. Europe mostly uses Brent crude which is extracted from North Sea oilfield and is light crude oil.

Besides being used as a fuel after processing, crude oil is essential raw material for manufacturing plastics, cosmetics and medicines. Goldman Sachs (US investment bank) estimates the proportion of crude oil used for primary materials production to be 45 percent.

Globally there are different types of crude oil, each having different property and pricing. For international trading purposes on futures exchange markets in London or New York, reference oils are used. These are standardized for determining the price of all other types of crudes. Some of them are:

- West Texas intermediate (WTI)
- Dubai Fateh
- Leona

- Alaska North Slope
- Urals
- Tijuana

International Petroleum Exchange in London is the most important trading venue for European Brent Crude Oil.

The pricing of crude oil varies widely depending on the quality of crude oil. In general, light sweet crudes carry a premium relative to medium and heavy sour crudes. The main reason for this variation being:-

- Light crudes have components that give more valuable lighter products
- Sweet crudes have less sulfur
- Light crudes require less energy to process
- The capital investment in equipment is less for processing of sweet crudes

The refiners (refining companies) face a big challenge in deciding between light sweet and sour crude for processing as light sweet crude is costly and gives better high-quality yield, on the contrary sour crude is cheap but needs heavy investment in processing equipment & high refining cost.

The economic and technical factors play an important role in pricing:

a. Crude quality differential pricing
b. Supply demand balance of crude
c. Local products in market
d. Product specification requirements
e. Local refining capacity
f. Upgrading capabilities
g. Global supply chain scenario affecting world oil price level

In general, the sweet crudes demand a premium of 15 to 25 % over the price of sour crude.

Chapter Sixteen

TAXATION ON PETROLEUM PRODUCTS

The Petroleum products are a big source of revenue for the Government. The Central government in India imposes various taxes. It can be understood easily by the fact that when Crude oil was USD 30 per barrel, then also there was no relief to the common man in pricing of Petroleum products. In year 2022, when the Crude oil is costing more than USD 100 per barrel, then also there is no relief given by government in pricing.

International gas and gasoline taxation encompasses a variety of concerns, including excise taxes, value-added taxes (VAT), and other charges imposed by other countries. Both the United States and the United Kingdom have separate tax schemes for petrol and gasoline, reflecting their own domestic policies and revenue requirements.

United States (USA):

1. Federal Excise Tax:

In the United States, there is a federal excise tax on gasoline. The federal excise tax on gasoline is 18.4 cents per gallon, and diesel fuel is 24.4 cents per gallon, according to my most recent knowledge update in January 2022. These levies help to pay federal roadway and transportation projects.

2. State Taxes:

Additionally, each state sets its own taxes on petrol and gasoline, resulting in price differences across states. Excise taxes, sales taxes, and other fees are examples of state taxes.

3. LOCAL TAXES: Some municipal governments may additionally collect additional gasoline taxes, adding to the overall taxation structure.

United Kingdom (UK):

1. Fuel Duty:

In the United Kingdom, gasoline and diesel are taxed. As of my most recent knowledge update, the fuel duty rate for gasoline is 57.95 pence per litre, and the same for diesel. This tax accounts for a large portion of the retail price of gasoline.

2. Value-Added Tax (VAT):

The United Kingdom levies a Value-Added Tax (VAT) on goods and services such as gasoline and diesel. The usual VAT rate is 20%, and it is applied to the total price, which includes gasoline duty.

3. Carbon Price Floor:

The United Kingdom has imposed a Carbon Price Floor, which is a surcharge on fossil fuels such as gasoline. This is intended to encourage a shift to cleaner energy sources.

International Considerations:

International petroleum fuel taxation issues entail a complex combination of economic, environmental, and geopolitical factors. Consider the following additional factors:

1. Cross-Border Fuel Pricing:

Geopolitical events, manufacturing costs, and taxation all have an impact on international fuel prices. Exchange rate fluctuations can also affect the cost of imported fuels.

2. Harmonization Efforts:

Some worldwide initiatives aim to harmonize tax regulations or promote transparency in fuel pricing. However, creating consistency between countries remains difficult.

3. Environmental Taxes:

The United States, the United Kingdom, and other countries are increasingly incorporating environmental factors into their tax policy. Carbon pricing, for example, aims to address the environmental consequences of gasoline consumption.

It's important to note that tax policies can evolve, and rates may change over time. Additionally, developments in international relations and environmental priorities can influence taxation strategies. For the most up-to-date and accurate information, it is recommended to refer to official government sources and tax authorities in each country.

A TYPICAL EXAMPLE OF TAXATION OF PETROL/DIESEL IN INDIA FOR YEAR 2022 IS GIVEN FOR ILLUSTRATION PUPOSES ONLY:

Sr. No.	Item of Taxation	Tax on Petrol (INR/Liter.)	Tax on Diesel (INR/Liter.)
1	Basic Excise duty	1.50	9.80
2	Special additional Excise duty	11.00	12.00
3	Agriculture tax, Road & Infrastructure Cess	15.50	
4	VAT (value added tax)	As per state govts.	As per state govts.
5	Additional local tax	As per state govts.	As per state govts.

Here it is important to note that the Central government passes only 42% of Excise duty received to the State governments. The state governments are not given any % of the Cess collected shown in above table. Additional taxes like value added tax (VAT) and local taxes are also added to the price by Local state governments as per their policy.

AT A GLANCE

Chapter: Seventeen

EMERGING TECHNOLOGIES AND FUTURE TRENDS

Crude oil and petroleum refining are critical components of the global energy landscape, supplying the fuels and raw materials required for contemporary living. As the globe grapples with issues such as energy security, environmental sustainability, and changing consumer demands, the refining industry is at the forefront of embracing new technology and creating future trends. This investigation digs into the cutting-edge technology and foresight trends that are set to transform crude oil and petroleum refining in the future years.

1. Digitalization and Industry:

Overview:

Digitalization is a transformative trend in the refining industry that uses advanced technologies such as the Internet of Things (IoT), artificial intelligence (AI), and big data analytics to improve efficiency, optimize processes, and enable predictive maintenance.

Key Technologies:

1. **IoT Sensors:** Throughout the refinery, connected sensors capture real-time data on equipment operation, temperature, pressure, and other critical characteristics.

2. **Big Data Analytics:** Advanced analytics analyzes massive volumes of data collected by sensors, providing insights into process improvement, supply chain management, and predictive maintenance requirements.

3. **Artificial Intelligence:** Systems using AI examine data trends, spot irregularities, and improve refining procedures to increase productivity and reduce energy use.

Benefits:
- **Predictive Maintenance:** The implementation of predictive maintenance can minimize maintenance expenses and downtime by identifying equipment problems early.
- **Energy Efficiency:** AI-driven optimization reduces environmental impact while increasing energy efficiency.
- **Process Optimization:** With the help of real-time data, refineries can modify their operations to maximize efficiency.

2. Hydrogen as a Clean Energy Carrier:

Overview:
Hydrogen is becoming more and more popular as a clean energy source and as a factor that could revolutionize the refining process. Its uses range from cutting emissions during the refining process to providing clean fuel for vehicles.

Key Technologies:
1. **Hydrogen Production:** Hydrogen is efficiently produced by using advanced technologies such as steam methane reforming (SMR) and electrolysis.
2. **Hydrogen Fuel Cells:** Hydrogen fuel cells provide a clean power source substitute by minimizing the need for conventional combustion methods.

Benefits:
- **Emission Reduction:** In the refining business, using hydrogen as a fuel or feedstock has the potential to drastically cut carbon emissions.

- **Decarbonization:** Hydrogen has the potential to aid in the Decarbonization of multiple industries, such as transportation and industrial operations.

3. Electrification of Refinery Operations:

Overview:
Electrification is the process of substituting electric-powered technologies for conventional combustion-driven operations, providing a more effective and environmentally friendly option.

Key Technologies:

1. **Electric Heaters:** Certain refining processes employ electrified heaters instead of conventional burned heaters, which lowers greenhouse gas emissions.
2. **Electric Motors**: Pumps, compressors, and other equipment are using electric motors more frequently, which improves energy efficiency.

Benefits:

- **Carbon Footprint Reduction:** Reliance on fossil fuels is reduced with electrification, which lowers carbon emissions.
- **Operational Efficiency:** Equipment running on electricity frequently has higher dependability and efficiency.

4. Bio-Based Feed stocks and Renewable Resources:

Overview:

As the industry looks for sustainable substitutes for conventional crude oil-derived inputs, the move towards bio-based feed stocks and renewable resources is gaining traction.

Key Technologies:

1. **Bio-Based Feedstock Processing:** Biofuels and biochemical are produced by technologies like bio-refineries and bio-crackers, which process renewable feed stocks like biomass.
2. **Algae-Based Technologies:** Algae's fast growth and capacity to sequester carbon make them a promising source for biofuels and other high-value products.

Benefits:
- **Sustainability:** Compared to traditional fossil fuels, bio-based feed stocks provide a more renewable and sustainable energy source.
- **Circular Economy:** The use of agricultural byproducts and waste streams lessens the need on finite resources.

5. Carbon Capture, Utilization, and Storage (CCUS):

Overview:

CCUS technologies aim to capture carbon dioxide emissions from industrial operations such as refining and either store or use them to reduce environmental effect.

Key Technologies:
1. **Carbon Capture Systems:** This Technology absorbs CO_2 emissions from a variety of sources, including refinery flue gases.
2. **Carbon Utilization:** Captured CO_2 can be used in procedures like enhanced oil recovery (EOR), or it can be transformed into valuable goods.
3. **Carbon Storage:** CO_2 may be safely kept underground, keeping it from being released into the environment.

Benefits:
- **Emission Reduction:** CCUS technologies help to reduce greenhouse gas emissions from refineries.
- **Resourceful Utilization:** Captured carbon can be used to improve oil recovery or to make chemicals and materials.

6. Advanced Catalysts and Materials:

Overview:
Technological developments in materials and catalysts are essential for increasing the sustainability, efficiency, and selectivity of refining operations.

Key Technologies:
1. **Nanotechnology**: Because of their increased catalytic activity, nanomaterial enable more effective and focused chemical reactions.
2. **Zeolite Catalysts:** Zeolites with specific compositions and structures improve the effectiveness of different refining procedures.
3. **High-Entropy Alloys**: Innovative alloys with distinctive compositions are being investigated for better corrosion resistance and equipment longevity in refining.

Benefits:
- **Process Efficiency:** Improved catalysts allow feed stocks to be converted into valuable products more effectively.
- **Durability and Sustainability:** Use of advanced materials have led to longer equipment lifespans and lower environmental impact.

7. 3D Printing in Refinery Construction:

Overview:
Additive manufacturing, or 3D printing, is transforming the way equipment and parts for refineries are built. With the use of this technology, complex structures can be created more effectively.

Key Technologies:
1. **Additive Manufacturing of Parts:** Pumps, heat exchangers, and valves are among the parts that can be made for refineries using precision 3D printing.
2. **Metal Additive Manufacturing:** Modern metal 3D printing technology makes it possible to produce intricate and long-lasting metal components for vital refinery machinery.

Benefits:
- **Customization:** Components can be customized with 3D printing, allowing designs to be optimized for particular refinery requirements.
- **Reduced Lead Times:** Complex part production lead times can be greatly shortened with additive manufacturing, accelerating project completion.

8. Artificial Intelligence for Operations and Optimization:

Overview:

Refining operations are rapidly incorporating artificial intelligence (AI) for predictive maintenance, process optimization, and real-time monitoring.

Key Technologies:
1. **Predictive Analytics:** Artificial intelligence systems examine both real-time and historical data to forecast equipment faults and suggest preventive maintenance.
2. **Machine Learning:** Machine learning algorithms increase operating efficiency and adjust to changing conditions to optimize refining processes.
3. **Digital Twins:** Operations in refineries are simulated, tracked, and predicted using digital twins-virtual copies of physical assets.

Benefits:
- **Operational Efficiency:** AI-driven optimization raises overall operational performance, decreases downtime, and increases efficiency.
- **Cost Reduction:** Through better resource use, predictive maintenance and process optimization reduce costs.

9. Circular Economy and Waste Valorization:
Overview:
Reducing waste, increasing resource efficiency, and obtaining value from waste streams and byproducts produced throughout the refining process are the main objectives of the circular economy strategy.

Key Technologies:
1. **Waste-to-Energy Conversion:** Technologies turn refinery waste into energy, lowering waste disposal's environmental impact.
2. **Resource Recovery:** Valorization techniques recover valuable materials or chemicals from byproducts of refineries, resulting in a closed-loop system.

Benefits:
- **Waste Reduction:** Circular economy strategies reduce waste output, which contributes to ecologically friendly and sustainable operations.
- **Resource Efficiency:** Valorization methods offer value to waste streams, increasing resource efficiency.

10. Advanced Monitoring and Sensors:

Overview:
Advanced monitoring and sensor technologies are critical for collecting real-time data and assuring the safety, efficiency, and dependability of refining processes.

Key Technologies:
1. **Wireless Sensor Networks:** Wireless sensors monitor the state of equipment and provide real-time data for preventive maintenance and operational optimization.
2. **Advanced Process Control:** Advanced control systems use sensor data to enhance refining processes and adjust to dynamic conditions.

Benefits:
- **Safety:** Continuous monitoring improves safety by recognizing anomalies and potential threats.
- **Efficiency:** Real-time data allows operators to make educated decisions while improving operations for efficiency and performance.

11. Enhanced Safety and Reliability with Robotics and Automation:

Overview:
Automation and robotics are becoming important in refining processes, leading to increased safety, dependability, and efficiency. Autonomous systems and robotics are used for a variety of jobs, ranging from basic inspections to sophisticated maintenance processes.

Key Technologies:
1. **Drones and UAVs:** For aerial inspections of refinery infrastructure, unmanned aerial vehicles (UAVs) equipped with sensors and cameras are used, providing a safer and more efficient alternative to manual inspections.
2. **Robotics for Maintenance:** Robotic arms and autonomous robots are used for maintenance operations in difficult or hazardous environments, minimizing the requirement for human interaction.
3. **Automated Monitoring Systems:** Continuous monitoring systems that include automatic alerts and reactions improve safety by detecting irregularities and potential problems in real time.

Benefits:
- **Safety:** Robotics and automation reduce human workers' exposure to hazardous circumstances, contributing to overall safety improvement.
- **Efficiency:** Automated systems can perform inspections and maintenance chores more quickly and precisely, saving downtime.

12. Sustainable Aviation Fuels (SAFs) and Alternative Feed stocks:

Overview:
As the aviation industry seeks to reduce its carbon footprint, the development and adoption of Sustainable Aviation Fuels (SAFs) are gaining prominence. SAFs are produced from renewable feed stocks, offering a cleaner alternative to traditional aviation fuels.

Key Technologies:

1. **Bio refineries for SAF Production:** Bio refineries use sustainable feed stocks such as waste oils, fats, or biomass to generate SAFs using procedures such as hydro processing or Fischer-Tropsch synthesis.
2. **Advanced Biofuel Technologies:** Innovative technologies, such as synthetic biology and metabolic engineering, aim to improve biofuel production efficiency and explore new feedstocks.
3. **Benefits:**
- **Carbon Reduction:** SAFs help to reduce carbon emissions in the aviation sector, which aligns with worldwide efforts to combat climate change.
- **Diversification of Feedstocks:** Investigating substitute feedstocks make the aviation fuel supply chain more diversified and sustainable.

13. Resilience and Risk Management with Predictive Analytics:

Overview:
Refineries are utilizing machine learning and artificial intelligence to fuel predictive analytics for resilience planning and risk management. These technologies use real-time and historical data analysis to forecast and reduce future dangers.

Key Technologies:

1. **Risk Prediction Models:** Algorithms using machine learning examine past data to spot trends and forecast possible dangers like supply chain interruptions or equipment breakdowns.
2. **Simulation and Scenario Analysis:** Refineries can carry out scenario assessments and assess the possible effects of different risk variables on operations thanks to sophisticated simulation tools.
3. **Integrated Risk Platforms:** Platforms that combine information from multiple sources offer a comprehensive picture of hazards, facilitating better decision-making.

EMERGING TECHNOLOGIES AND FUTURE TRENDS

Benefits:
- **Preventive Maintenance:** Refineries can reduce unplanned downtime and enable preventive maintenance by using predictive analytics to forecast equipment faults.
- **Operational Continuity:** Predictive analytics-based risk management techniques improve the overall resilience of refining operations.

14. 5G Connectivity for Real-time Communication and Control:

Overview:
Refinery control and communication systems are undergoing a revolution thanks to the introduction of 5G technology. Real-time data exchange, control, and monitoring are made possible by high-speed, low-latency connectivity, which makes operations more responsive and effective.

Key Technologies:
1. **IoT Integration:** A large network of IoT devices is supported by 5G connectivity, enabling real-time equipment, asset, and process monitoring.
2. **Augmented Reality (AR) for Maintenance:** 5G-enabled augmented reality apps let remote specialists mentor on-site staff members during maintenance tasks, increasing productivity and decreasing downtime.
3. **Wireless Control Systems:** Refinery process control can be made more responsive and nimble thanks to 5G's ability to support wireless control systems.

Benefits:
- **Real-time Decision Making:** Minimal latency Real-time data transmission and reception is ensured by 5G connectivity, facilitating prompt decision-making.
- **Remote Operations:** Refinery operations may be remotely monitored and controlled thanks to 5G connectivity, which eliminates the need for on-site personnel.

15. Regulatory Compliance and ESG Reporting:

Overview:
Environmental, social, and governance (ESG) reporting and regulatory compliance are becoming more and more important for refining operations. Cutting edge technology are being used to make sure that rules are followed and that the effects on the environment and society are reported openly.

Chapter – Eighteen

INTERNATIONAL SCENARIO (2022-23)

The Ukraine war brought chaos and drastic changes in the international energy market. Russia's invasion in Ukraine brought fears of an energy crisis in Europe.

Initially crude oil prices soared spirally due to Russia slashing supplies, but were brought in control by some sanctions raised by Western countries on Russia to reduce revenues. Europe also made alternate arrangements reducing their dependence on Russian crude.

The G-7 group of countries (US, Canada, UK, France, Italy, Germany & Japan) capped the Russian crude oil price to $60 a barrel on 5th December 2022. The European Union also implemented the same in February'2023.

As a result, the Russian Urals crude oil prices have slipped from $95 (on the day of Ukraine invasion i.e., Feb-2022) to $60 a barrel (as of Feb 2023). At one instance in Jan-2023 the Ural crude traded at $50 a barrel also.

Non G7 countries like India and China have taken the opportunity of buying cheap fuel from Russia in large quantities. G-7 was earlier called G-8 until Russia was expelled from it.

According to recent data (source: Kpler, commodities market data), China is buying 1.7 million barrels of Ural crude per day. India is also importing 1.2 million barrels of crude oil per day from Russia.

NATURAL GAS

Countries all around the world are depending more and more on natural gas as a clean and effective energy source in the global natural gas situation. Natural gas is a flexible fuel used for heating, power generation, and industrial activities. It is mostly made of methane. Because it releases fewer toxins when burned, it is thought to be cleaner than many conventional fossil fuels.

To enable the transportation of natural gas over international borders, nations are investing in infrastructure related to this fuel, including pipelines and facilities for LNG. As a result, a worldwide market for natural gas has developed, with nations purchasing and selling gas to meet their energy demands.

Due to its lower carbon emissions than coal and oil, natural gas is becoming more and more popular as a transitional energy source in the worldwide effort to combat climate change. Furthermore, natural gas facilitates the switch to renewable energy sources by offering a dependable and adaptable energy source.

Natural gas is essential for striking a worldwide balance between energy security, economic growth, and environmental objectives, particularly in light of the growing emphasis on sustainability in the global society. The goal of this cooperative strategy is to provide nations all around the world with cleaner, more sustainable energy in the future.

In order to cripple Europe's economy, Russia has totally cut off Natural gas supplies through Nord Stream 1 pipeline to Europe. Other countries have responded to this and Australia, Qatar and USA have augmented LNG (liquefied natural gas) gas supplies to Europe to cope up with its demand.

Infrastructure and energy supply can generally be disrupted by wars and geopolitical unrest. Since Europe depends largely on Russian natural gas imports, any disruptions in Russia's production or transportation capacity could have an impact on the country's supply of natural gas to other regions. Russia is a major exporter of natural gas.

In the context of the conflict between Russia and Ukraine, some factors that can have an impact on the supply of natural gas include:

Pipeline disruptions: The supply chain may be impacted by any harm or interruptions to the pipelines that carry natural gas from Russia to Europe.

Geopolitical conflicts: These conflicts have the potential to impact trade relations and investment decisions.

Chapter Nineteen

BRIEF SUMMARY

BASIC PROCESSING OPERATIONS

The basic petroleum processing operations can be classified into 7 categories:

1. SEPARATION: using distillation, crystallization, absorption and solvent extraction

2. CONVERSION: removal of carbon and addition of hydrogen

3. REFORMING: cracking, hydrocracking, visbreaking, coking and reforming

4. REARRANGEMENT: isomerization

5. COMBINATION: alkylation & catalytic polymerization

6. TREATING EMULSIFYING & BLENDING: to produce gasoline, diesel, kerosene, lubricants, wax, asphalt (bitumen)

7. ENVIRONMENT PROTECTION: sulfur recovery, waste water treatment and disposal of solids

PHYSICAL & CHEMICAL PROCESSES

The various physical & chemical processes used are summarized below:

PHYSICAL PROCESSES:

- Distillation
- Solvent extraction
- DE asphalting
- Solvent dewaxing
- Blending

CHEMICAL PROCESSES (THERMAL):

- Visbreaking
- Coking

CHEMICAL PROCESSES (CATALYTIC):

- Alkylation
- Polymerization
- Isomerization
- Catalytic reforming
- Hydro treating
- Catalytic cracking
- Hydrocracking
- Catalytic dewaxing

*****

ABOUT THE AUTHOR

In the realm of self-growth and safety, P K Singh emerges as a dedicated explorer and advocate, navigating the intricate landscapes of personal development and well-being.

Armed with academic qualifications and experience in the Energy sector, the author is an Electrical Engineer from Delhi University and a certified Trainer from the Institute of Learning & Management, UK. He is a DNV, and Bureau Veritas certified ISO 9001-2015 QMS/EMS Lead Auditor having more than 40 years of experience in handling SHE, Operations, Logistics, HR, Recruitment, and Training functions in various multinational industries in India and abroad.

As an author, he has translated this wealth of knowledge into the written word, offering readers a road-map for personal development and strategies to enhance their safety and well-being. P K Singh's books are characterized by a blend of research-backed insights, practical advice, and a compassionate understanding of the challenges individuals face on their journey toward growth and self-discovery.

Through the pages of his Books, he invites readers to embark on a transformative journey, fostering a deeper connection with oneself and cultivating an environment of security and well-being.

Thank you

A
Book from
"pk resources"
India

www.ingramcontent.com/pod-product-compliance
Lightning Source LLC
Chambersburg PA
CBHW052352220526
45465CB00003BA/1077